JOURNAL OF SELF-ASSEMBLY AND MOLECULAR ELECTRONICS

Volume 1, No. 2 (May 2013)

JOURNAL OF SELF-ASSEMBLY AND MOLECULAR ELECTRONICS

Editors-in-Chief

Peter Fojan and Leonid Gurevich
Department of Physics and Nanotechnology
Aalborg University
Denmark

Aims

Self-Assembly and Molecular Electronics (SAME) is a multidisciplinary peer-reviewed journal with a wide-ranging coverage, specializing in the areas of molecular electronics and self assembly systems. SAME encourages original cross-disciplinary full research articles, rapid communications of important scientific and technological findings, and state-of-the-art reviews.

Scope

SAME publishes theoretical and experimental original research covering areas of:

- Molecular Electronics and Molecular Devices with a particular emphasis on DNA, peptide and protein based systems
- Self Assembly in Nanosience, Chemistry, Biology and Medicine
- Supramolecular Chemistry
- Modelling of Structural and Electronical Properties of Organic Molecules and Self Assembled Systems

Published, sold and distributed by:
River Publishers
Niels Jernes Vej 10
9220 Aalborg Ø
Denmark

Tel.: +45369953197
www.riverpublishers.com

Journal of Self-Assemly and Molecular is published three times a year.
Publication programme, 2013: Volume 1 (3 issues)

ISSN 2245-4551
ISBN 978-87-93102-39-2 (this issue)

JOURNAL OF SELF-ASSEMBLY AND MOLECULAR ELECTRONICS

Volume 1 No. 2 May 2013

Supramolecular Organization of Bimetallic Building Blocks: from Structural Divertimentos to Potential Applications

Ruben Mas-Balleste* and Felix Zamora*

Departamento de Química Inorgánica, Universidad Autónoma de Madrid, 28049 Madrid, Spain
Corresponding authors: ruben.mas@uam.es; felix.zamora@uam.es

Received 23 April 2013; Accepted 22 May 2013; Publication 6 June 2013

Abstract

The phenomena that results on supramolecular aggregations of bimetallic $[Pt_2L_4]$ and $[AuL_2]$ are reviewed. Supramolecular $[AuL_2]_n$ (n = 2,3) were observed in some cases in solution as a result of Au(I)···Au(I) aurophilic interactions, which also direct the assembly of oligomeric structures in crystal phase. Analogously, Pt(II)···Pt(II) attraction accounts for the assembly of $[Pt_2L_4]_n$ supramolecules which can result on 1D semiconductive arrangements in crystal phase and direct the formation of 1D nanofibres on surfaces. Finally, oxidation of $[Pt_2L_4]$ to $[Pt_2L_4I]_n$ produces highly conductive polymers that can reversibly assemble/disassemble into $[Pt_2L_4]$ and $[Pt_2L_4I_2]$. Such outstanding ability results on an unprecedented processability that enables MMX polymers for technological applications as molecular wires.

Keywords: Metal-metal interactions, Supramolecular assemblies, MMX Chains, Molecular wires.

1 Introduction

In recent years, considerable attention has been paid to the design and synthesis of molecular materials and devices [1] being the organization of molecules a main goal in both material science and nanoscience. The systematic use of

Journal of Self-Assembly and Molecular Electronics, Vol. 1, 149–176.
doi: 10.13052/jsame2245-4551.121

weak interactions to assemble complex structures is the ultimate challenge in supramolecular chemistry.[2] But while hydrogen bonding has been widely studied, other types of intermolecular interactions such as weak metal···metal interactions still remain less explored.[3] On the other hand, compounds containing metal-metal interactions is a field that has attracted great attention during years in material science because of their unusual electronic properties, magnetism and/or electrical conductivity.[3b] This "old topic" has gained renewed attention in the preparation and characterization of materials with high anisotropy and restricted dimensionality. The reason is very simple and can be understood taking into account that small pieces of matter may show very unusual optical, magnetic, and conducting properties due to their size once they go into the nanoscale.

Metal-metal interaction is the driving force to form some linear structures with interesting properties. Several reports suggest that weak metal-metal interactions can be used to build up complex structures.[3b] Such attractive interactions are mainly observed between units containing Pt(II), Ag(I) or Au(I) centres. For example, tetracyanoplatinates, $K_2Pt(CN)_4X_{0.3}\cdot nH_2O$ (X = Cl, Br), named KCPs, were one of the first examples showing electrical conductivity.[4] These systems are formed by the stacking of square-planar $[Pt(CN)_4]^{n-}$ anion complexes.[5] The structures show the overlapping of $5d_z{}^2$ orbitals strongly affected by the Pt-Pt intermolecular distances that also affect to their conductive properties. KCP based materials represent the first inorganic "molecular wires" ever designed.[6] Similarly, analogous columnar systems have been formed with dimetallic precursors, for instance $[Pt_2(S_2CR)_4]$ (R = alkyl group).[7] These discrete dinuclear complexes with intermolecular metal-metal interactions have fascinating magnetic and electrical properties.[8] Some of them have been shown to be suitable as precursors for weakly-bound one-dimensional metal chains.[9] But probably one of the most studied metal-metal interactions is the tendency of Au(I) centres to establish weak Au(I)···Au(I) bonding, named aurophilia.[10] This phenomenon is a consequence of a combination of dispersion forces and the mixing of filled 5d-based molecular orbitals and empty molecular orbitals of appropriate symmetry derived from the 6s and 6p orbitals.[11] As a consequence of this supramolecular interaction, and taking advantage of the structural versatility of Au(I) centres, a variety of structures with interesting optical and/or electrical properties have been reported.[12, 13]

However, a deeper understanding of the chemical principles that direct such assemblies is still an important challenge in chemistry and could probably has an impact in nanoscience.

The goal of this review is to provide a current overview on the recent studies devoted to bimetallic species containing the basic structural units $[Au_2(S_2CR)_2]$ or $[Pt_2(S_2CR)_4]$ as well as the linear structures resulting from combination of diplatinum units and iodide bridging ligands.

2 Gold(I)···gold(I): Golden attraction

2.1 In Crystal Phase

Crystal structures reported for homoleptic Au(I)-dithiocarboxylato compounds reveal an outstanding ability to generate oligomeric/polymeric structures by means of the μ-kS:kS' bridging mode. From such structures a complex series of intra- and inter-molecular supramolecular interactions can be found. Only in the case of $[Au_2(\text{isobutyldithiocarboxylato})_2]$, the existence in the crystalline phase of two polymorphic forms have been reported: one formed by discrete $[Au_2L_2]$ bimetallic units and other formed polymeric $[AuL]_n$ chains. In that case, intermolecular Au···Au distance all fall far from the range of significant aurophilic interactions.[14]

The scenario found for xanthato ligands differs substantially from what is observed for dithiocarboxylato containing compounds. All structures known consist on discrete dimeric entities in which the metal centres are joined by double μ-kS:kS' xanthate bridges leading to essentially planar eight-member $Au_2S_4C_2$ rings. While strong intramolecular aurophilic interaction between the gold(I) centres is generally observed, not allways significant intermolecular interactions are found. Only in the case of $[Au_2(\text{isopropyl-xanthate})_2]$ Au···Au at around of 3.0 Å have been observed indicating aurophilic attractive intermolecular interactions.

The preponderance of double bridged dimers observed for xanthato ligands is also found for dithiocarbamate containing Au(I) compounds. However, dithiocarbamate containing $[Au_2L_2]$ structures tend to establish short Au··Au intermolecular distances. For the case of the dithiocarbamate, there is one ligand that provides both single-bridge-based (oligomeric) structures and double bridged dimers. However, in that case, only a severe increase of the concentration of gold in the reaction media, allows to overcome the appearance of the more favourable dimeric-based crystal structure.

Overall, as shown in Figure 1 regarding the nuclearity of crystallographically characterized $[Au_2L_2]$ compounds, it appears that there is a higher tendency of the dithiocarboxylato ligands to provide single-bridge-based structures (rings and chains) than in the case of xanthate and dithiocarbamate

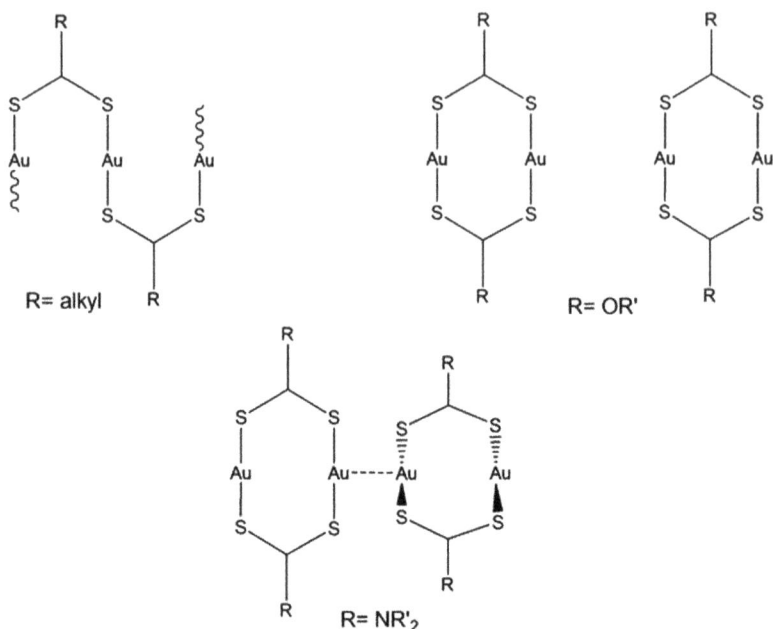

Figure 1   Preferred structural motifs for homoleptic Au(I) compounds containing dithiocarboxylato, xanthato or dithiocarbamato ligands.

ligands, where double bridged dimers are preponderant. While xanthates prefer do not establish intermolecular aurophilic interactions, dithiocarbamate containing complexes follow the opposite trend.

Supramolecular arrangements based on [Au$_2$L$_2$] units can mainly be attributed to an Au···Au attraction known as aurophilia. While the most stabilizing intermolecular interaction is generally that between two gold atoms, also Au···S and S···S interactions have been described as driving forces for such assemblies.[14–15] Typically, when the Au···Au is found at around 3Å the intermolecular interaction between Au(I) centres is maximized. Interestingly, the Au···S interactions are also stabilizing but at somewhat longer distances (with a minimum at around 3.5 Å), whereas S···S interactions are destabilizing at short distances and only slightly stabilizing at 3.8–4.0 Å. Analysis of the crystal structures published for [Au$_2$L$_2$] units (being L dithiocarboxylates, xanthates, or dithiocarbamates) shows the interplay of the three intermolecular interactions and of their different distance dependence. Regarding the relative orientation of the dinuclear [Au$_2$L$_2$], has been found that Au···Au distances shorter than 3.30 Å are only compatible with a rotated structures, *i.e.* with

S-Au⋯Au-S torsion angles larger than 30°. In contrast, coplanar orientations are only compatible with Au⋯Au distances larger than 3.5 Å. This effect can be clearly appreciated in Figure 2, where a representation of the interdimer Au⋯Au distances (from a CSD search for compounds [Au_2L_2], $L = {}_2$SC-X-R, X = C,N,O) as a function of the S-Au⋯Au-S torsion angle is shown.

In addition, another geometrical parameter that serves to evaluate the total interaction energy is the relative orientation of the interacting AuS_2 coordination moieties, defined by the Au-Au⋯Au angles (ω). An analysis of all short contacts between those units in the CSD shows that the Au⋯Au distances are all longer than 3.88 Å when a gold is found at a ω angle higher than 30°, suggesting that the interaction energy is most stabilizing at the perpendicular orientation.

In general, DFT performs poorly at describing accurately the energetics of Au(I)⋯Au(I) interactions. Thus, in order to rationalize the variety of structural patterns found for the supramolecular arrangement of the di- and oligonuclear Au(I) species, computational studies were carried out at the MP2 level of theory, capable of reasonably evaluating the relevant intermolecular interactions. Angular dependence of the strength of the aurophilic interaction was confirmed by BSSE-corrected interaction energy calculations between two dinuclear molecules at a fixed intermolecular Au⋯Au distance of 3.0Å and varying Au-Au⋯Au angles (ω). Theoretical results show that the interaction energy

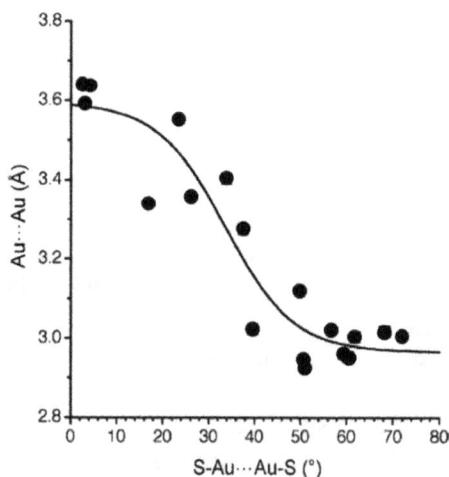

Figure 2 Dependence of the intermolecular Au⋯Au distances on the S-Au⋯Au-S torsion angle.

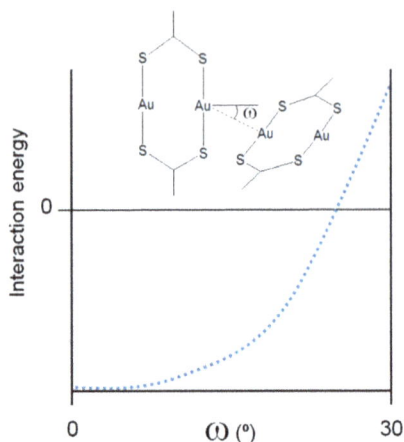

Figure 3 Theoretically computed dependence on the Au-Au···Au angles of the interaction energy between two [Au$_2$L$_2$] units.

varies little for deviations of less than 10° from a perpendicular arrangement, but become much less stabilizing at larger angles (Figure 3).

Optimization of the structures [Au$_2$L$_2$]$_2$ (being L dithiocarboxylato or xanthato ligands, see Figure 4), illustrated these concepts. The most stable geometries with each ligand, present structures with stacked Au$_2$S$_4$C$_2$ rings, shifted along the Au–Au direction to form Au4 rhombuses, and also rotated relative to each other. Such arrangement facilitates the existence of three intermolecular Au···Au contacts at distances of around 3.0 Å and four Au···S contacts at 3.5 Å or shorter, whereas avoiding S···S short contacts.

2.2 In Solution

An intriguing feature of [Au$_2$L$_2$] (L = dithiocarboxylato) is that for a series of compounds for which different crystal structures have been found, an identical behaviour in solution (in CS$_2$) has been observed. In fact, the polymeric/oligomeric [Au$_2$L$_2$]$_n$ structures found in the crystal phase are cleaved to form discrete analogous [Au$_2$L$_2$] entities. Furthermore, spontaneous self-assembly that depends on the temperature and the concentration of the gold compound has been observed for such compounds. Variable temperature UV-vis data shows that lowering the temperature triggers the formation of a new feature at slightly lower energies with respect to that observed at room temperature. This observation was assigned to an aggregation process which reversed upon heating. The appearance of different species in the solutions of [Au$_2$L$_2$]

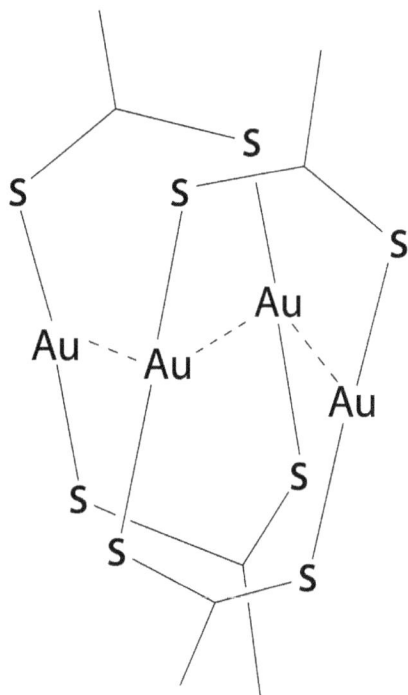

Figure 4 More stable configuration of the [AuL$_2$]$_2$ structure, theoretically optimized.

(L = dithiocarboxylato) by lowering the temperature was also observed by ^{1}H NMR spectroscopy. The assessment of the degree of association in solution was obtained by measuring diffusion coefficients (D) from DOSY spectra of a solution at 240 K. According to the measured diffusion coefficient values, the species with a higher hydrodynamic radius (i.e., lower D) become more favourable at lower temperatures. The values obtained in all cases were consistent with the presence of a dimetallic, a tetrametallic, and a hexametallic species [Au$_2$L$_2$]$_n$ (n = 1, 2, 3) (Figure 5).

A similar study was performed for analogous Au(I)–xantate compounds. Surprisingly, under the same reaction conditions, for compounds with xanthate ligands no spectroscopic changes (UV-vis and NMR) were found upon decreasing the temperature. Thus, although dithiocarboxylate-containing compounds spontaneously aggregate in solution, analogous complexes with xanthate ligands do not show such behaviour. Different behaviour in solution of [Au$_2$L$_2$] depending if L is dithiocarboxylato or xanthato constitutes a case of subtle modulation of the supramolecular interactions between

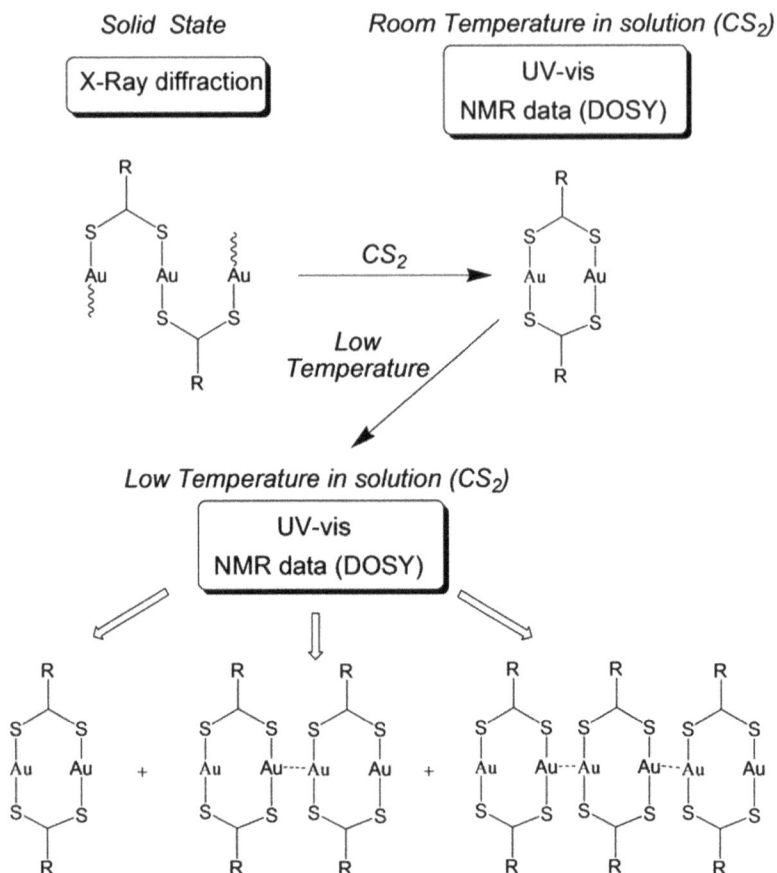

Figure 5   Behaviour in solution observed when $[Au_2L_2]n$ (L = dithiocarboxilato) are dissolved in CS_2.

compounds containing Au(I) centres. Considering that steric congestion is equivalent for analogous dithiocarboxylato and xanthato ligands, the tuning of aurophilic assembly should rely on electronic effects. The different behaviour of $[Au_2L_2]$ compounds has been attributed to the different electron-donor or electron-withdrawing nature of L ligands. For the case dithiocarboxylato containing compounds, both oligomeric/polymeric structures and the discrete dimeric ones are generated from the corresponding discrete dimeric $[Au_2L_2]$ entities present in solution. Thus, it is reasonable to assume that the formation of the oligomeric/polymeric $[Au_2L_2]_n$ structures take place through a previous intermolecular aggregation stage of the dimers mediated by the intermolecular aurophilic contacts. Thus, when aggregation in solution is observed (for the dithiocarboxylato compounds) high nuclearities have been found in the crystal phase.

3 Platinum(II)···Platinum(II): Platinum Weddings

3.1 In Crystal Phase

All the reported crystal structures for compounds of the formula $[Pt_2L_4]$ ($L = S_2CR$) show linear arrangements that are directed by intermolecular Pt(II)···Pt(II) interactions. There is only one exception to this tendency for a polymorph reported of $[Pt_2(S_2CCH_3)_4]$, which does not show short Pt···Pt distances. Typically, the crystal structures of such compounds consist of quasi-one-dimensional chains based on collinear alignment of $[Pt_2(S_2CR)_4]$ dinuclear entities with short intra- and interdimeric Pt–Pt distances (Figure 6). The dinuclear entities show a windmill-shaped arrangement in which four μ-dithiocarboxylato-kS:kS' ligands bridge two Pt(II) centres. The $[Pt_2L_4]$ complexes are stacked collinearly along the intradinuclear Pt–Pt axis by means of short interdimer Pt···Pt distances (3.0–3.4Å). The steric hindrance of the substituents of the dithiocarboxylato ligands influences the interdimeric Pt···Pt distances: the bulkier is the substituent, the longer is the interdimer Pt···Pt distance. This effect is reflected on the electrical properties measured on single crystals. Typically, the conductivity measured in single crystals is the characteristic of semiconducting materials and exhibits a remarkable dependence on the intermolecular metal-to-metal distances, being observed that short intermolecular Pt···Pt distances enhance conductivity.[9b, 16]

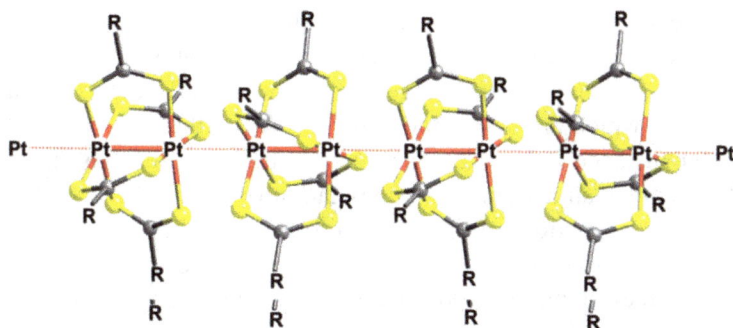

Figure 6 Supramolecular 1D arrangement typically observed in crystalline samples of $[Pt_2L_4]$ compounds.

In principle, analogously to what is observed for Au(I) compounds, one would expect some kind of attraction between Pt and S atoms of adjacent monomers. However, theoretical calculations, indicate that intermolecular S···S interaction is of repulsive character in $[Pt_2L_4]$. As a consequence of such repulsive interactions the energetic minimum adopts a structure with a staggered conformation with dihedral angle S-Pt···Pt-S values close to 45° in both crystallographic and calculated structures. These observations, emphasize that the stability of $[Pt_2L_4]_n$ supramolecular entities is due to persistent weak intermolecular metal···metal interactions, which are in general difficult to describe by standard DFT methods. In addition, according to theoretical calculations, relativistic and electronic correlation effects are thought to be important in the stability of such aggregates. Thus, to describe accurately the energetics of $[Pt_2L_4]$ dimerisation and predict Pt···Pt bond lengths comparable with the available crystallographic experimental data, it is necessary to use a complete basis set and exchange-correlation DFT functionals. Thus, the widely used exchange-correlation functionals BLYP and B3LYP predict a repulsive Pt···Pt interaction, a situation that even the long-range corrected functional CAM-B3LYP is not able to correct. After extensive testing, was found that the PBE functional family (which includes its hybrid version PBE0) in the complete basis set limit, provides an optimal compromise between accuracy and computing time.

Theoretical analysis of the charge distribution in $[Pt_2L_4]_2$ highlighted the leading role of the ligands L in the assembly of $[Pt_2L_4]_n$ through Pt···Pt interactions. Specifically, Natural Population Analysis shows the ability of the sulphur atoms to accommodate a part of the charge donated by the platinum atoms resulting in a synergic effect in the stable linear structures.

A first approach to understand the electronic description of the Pt···Pt bonding between monomer species, can be attempted by computing the change in the energy and the shape of the occupied orbitals depending on the Pt···Pt distance. The major changes are observed at the HOMO and HOMO-1 levels, which are degenerated up to 6 Å. Further decrease of the intermolecular distance induces an energy splitting of these MOs. In fact, the HOMO becomes an antibonding combination of mainly $5dz^2$ character with significant contributions of the 6pz and 6s orbitals of Pt and its energy increases upon dimer formation. In contrast, the HOMO-1 is of bonding nature and becomes stabilized upon dimerisation. From this description it is not inferred a neat stabilization, and thus is concluded that the combination of adjacent occupied Pt $5dz^2$ orbitals alone is not sufficient to explain the attractive interactions between Pt(II) centres. Thus, Pt(II)···Pt(II) attraction

should be explained according to more subtle interactions. The current understanding of the driving force of such weak metal–metal interaction involves a
symmetry-allowed mixing between atomic orbitals of adjacent Pt atoms along
the Pt–Pt vector namely, the occupied (donor) $5dz^2$ and the empty (acceptor)
6pz and 6s orbitals. Such donor–acceptor interaction is represented in terms
of atomic orbitals in Figure 7.[17]

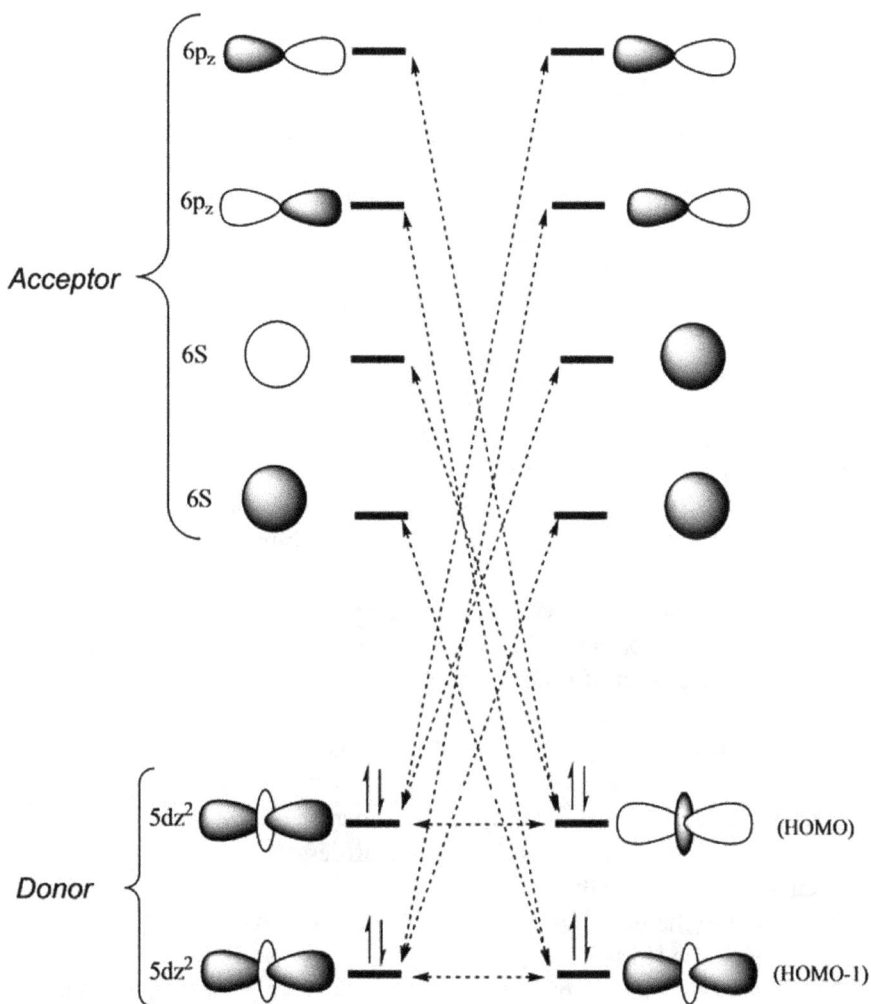

Figure 7   Orbital interactions that account for Pt(II)···Pt(II) supramolecular assembly.

3.2 In Solution

Similarly to what is observed for Au(I), compounds [Pt_2L_4] present a thermochromic behaviour, which depends on concentration, temperature, solvent and nature of ligand. Generally, by decreasing the temperature, a new broad band appears in the red/near-infrared region. This new spectroscopic feature is a very wide band that disappears upon heating and re-appears when the solution is cooled down again. The dependence upon the concentration of such absorbance indicates that it is due to reversible aggregation processes in solution. Determination of the degree of aggregation is in this case more controversial than for Au(I) compounds. A first proposal for aggregation of complex [$Pt_2(S_2C(CH_2)_5CH_3)_4$] was an equilibrium between the species [Pt_2L_4] and [Pt_2L_4]$_2$. However, higher nuclearities cannot be discarded.[18] In this case NMR data have not provided a definite answer. Variable-temperature (VT)-^{1}H NMR measurements did not allowed to distinguish signals from different oligomers. Thus, hints on the nature of supramolecular aggregates should be obtained from a detailed observation of electronic spectroscopy. For the case of complex [$Pt_2(S_2C(CH_2)_4CH_3)_4$], which presents a good solubility even at low temperatures, several observations indicate that a mixture of different nuclearities can be found in solution at low temperature.

A further support for the assignments of UV-vis spectra at low temperature was obtained from theoretical investigations of the spectroscopic properties of isolated [Pt_2L_4]$_n$ (n = 1–4) species (Figure 8). The calculated most intense absorption peaks for the different [Pt_2L_4]$_n$ species are found at approximately $\lambda = 410, 570, 686$ and 770 nm for the series n = 1–4. These peaks correspond to neutral excitations from the HOMO (mostly an σ^* combination of the Pt $5dz^2$ orbitals) to unoccupied orbitals of s character involving the sulphur ligands. The theoretical simulation of spectra of [Pt_2L_4]$_n$ (n = 1–4) clearly indicate that increasing the length of the oligomer results in a decrease of the energy of the optical transition associated to Pt$\cdots$Pt aggregations. Comparison of the experimental and computed spectra clearly suggests that at room temperature only monomeric [Pt_2L_4] species are present, whereas at low temperature a mixture of [Pt_2L_4]$_n$ species with different nuclearities coexists. Furthermore, these calculations show that the position of the band peak in the visible region is very sensitive to the intermolecular Pt$\cdots$Pt distance, which could partially contribute to the broadening of this band.[17]

Supramolecular assembly of compounds [Pt_2L_4] can be affected by subtle parameters such as the nature of the ligand L or the solvent in which is dissolved the diplatinum compound. Under the same conditions, there

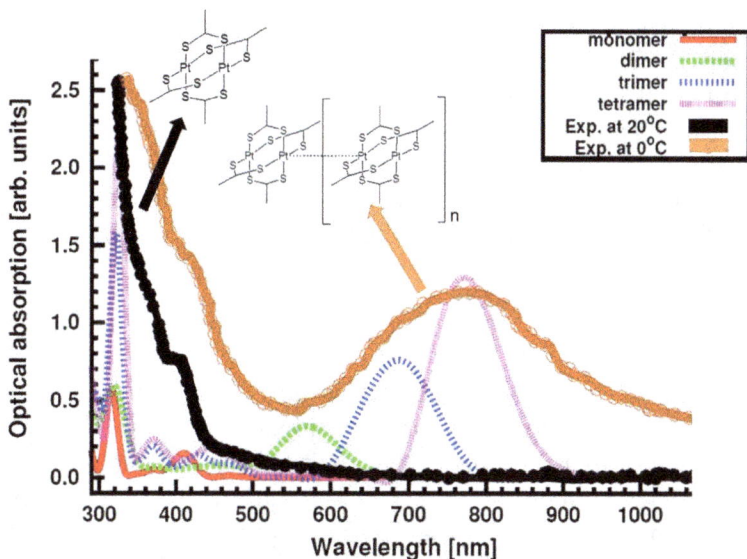

Figure 8 Experimental UV-vis spectra of $[Pt_2(S_2CH_3)_4]$ at 20°C or 0°C in CH_2Cl_2, together with the calculated spectra for $[Pt_2(S_2CH_3)_4]_n$ ($n = 1$–4).

is an evident effect of the ligand that can be related to the steric hindrance that each dithiocarboxylato ligand present in solution. Thus, for $L = CH_3\text{-}CS_2{}^-$, association is easier than with any other diplatinum complex. When the ligand contains a branched R group the approaching of the dimetal units is definitely more difficult than for compounds that contain linear R groups. Even small differences on the length of linear R chains can have an impact on supramolecular assembly of $[Pt_2L_4]$, being favoured when the R group is a shorter chain. However, crystal structures of $[Pt_2(S_2C(CH_2)_nCH_3)_4]$ ($n = 0$, 3 and 4) show similar intermolecular Pt···Pt distances. Thus, the effective bulkiness of the dithiocarboxylato ligand is not the same in solution and in crystal phase. In fact, ordered packing minimizes steric repulsive interactions especially where linear alkyl chains are present in the structure. However, free motion of such chains in solution provokes significant hindrance, which increases with the length of the alkyl group. This concept is less applicable to branched chains which present a high steric hindrance in solution that cannot be as well overcome in the ordered packing of the crystalline samples.

The effect of the solvent in self-assembly of compounds $[Pt_2L_4]$ in solution has been examined by comparing their behaviour using CH_2Cl_2, $CHCl_3$, THF and CS_2 as solvent. Under the same conditions, whereas a certain degree of self-assembly in CH_2Cl_2 and (in less extent) in $CHCl_3$ is observed, CS_2 and THF hamper the association in solution. This effect could, in principle, be justified considering the known coordinative ability of THF and CS_2, which can result in weak metal–ligand interactions between the solvent molecules and the platinum centres. However more subtle effects should be invoked to explain the different effect of CH_2Cl_2 and $CHCl_3$, which are probably related to solvation effects.

3.3 From Solution to Surface

The self-assembly observed in solution at low temperature triggered the generation of nanofibers on surfaces (Figure 9). This effect was observed by comparing the results of the adsorption on surfaces of $[Pt_2L_4]$ ($L = CH_3-CS_2^-$) at low temperature and at room temperature. When a diluted solution of $[Pt_2(dta)_4]$ in CH_2Cl_2 was cooled down to -50°C and adsorbed, on cooled HOPG surface, AFM images show nanofibers of lengths of up to several microns and heights ranging from 4 to 7 nm. However when the same experiment was carried out at room temperature, only amorphous material on HOPG was observed. Interestingly, nanofibers were observed in HOPG but not in mica. Thus, organization of $[Pt_2L_4]$ somehow takes place on the surface, which plays a leading role on the self-assembly process. In particular, as previously observed on related 1D systems, diffusion of small metallic fragments is much easier on HOPG, which can make easier the assembly of nanofibers. Overall, oligomerization of $[Pt_2L_4]$ in a cold solution facilitates the growth of long 1D structures on the surface on an appropriate surface.[9a]

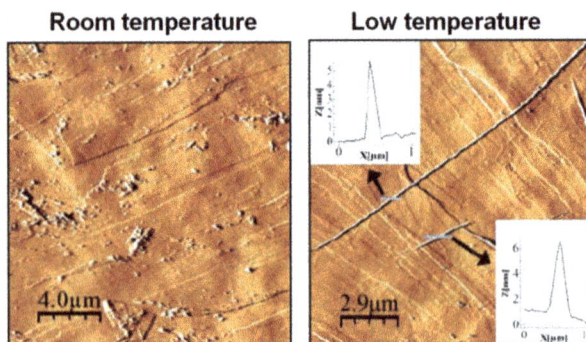

Figure 9 AFM images measured on HOPG surfaces after deposition of solutions of $[Pt_2(S_2CH_3)_4]$ in CH_2Cl_2 at 20°C and at -50°C.

4 MMX Chains: Donors and Acceptors with Ephemeral Affairs

4.1 In Crystal Phase

MMX chains belong to the family of mixed-valence metal-organic compounds.[19] Some of these metal-organic polymers show interestingphysical properties.[3] For instance, those based on ruthenium, dicarboxylate and halides, $[Ru_2(RCO_2)_4I]_n$ (R = alkyl group), have shown magnetic properties,[20] and those with platinum, dithiocarboxylate and iodine, $[Pt_2(RCS_2)_4I]_n$ (R = alkyl group), have shown metallic conductivity at room temperature (Figure 10).[21] Monocrystals of $[Pt_2(RCS_2)_4I]_n$ chains have attracted attention in view of their electrical conductivity at room temperature.[21–22]

Another structural feature of $[Pt_2(RCS_2)_4I]_n$ monocrystals is the different types of phase of extreme valence-ordering states that can show depending on the charge states of the metal atoms in the MMX chains [23]: (i) an averaged-valence state (AV): $M^{+2.5}$ - $M^{+2.5}$ — X — $M^{+2.5}$ - $M^{+2.5}$ — X , (ii) a charge polarization state (CP): M^{+2} - M^{+3} - X — M^{+2} - M^{+3} - X , (iii) a charge density wave state (CDW): M^{+2} - M^{+2} — X - M^{+3} - M^{+3} - X , and (iv) an alternate charge polarization state (ACP): M^{+2} - M^{+3} - X - M^{+3} - M^{+2} — X . The AV and CP states, in which the periodicity of the chain is M-M-X-, are consistent with a metallic and a Mott-Hubbard semiconducting state, respectively. The CDW and ACP states, described as Peierls and spin-Peierls states, are sensitive to structural parameters, and have a doubled M-M-X-M-M-X periodicity unit.

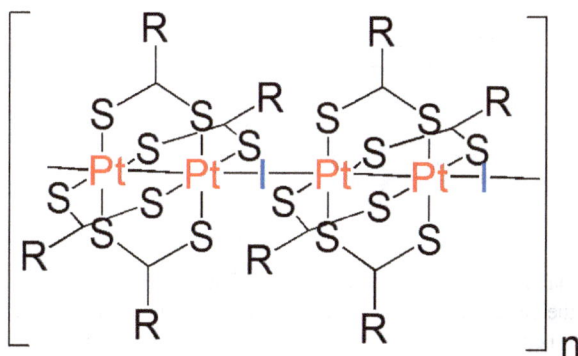

Figure 10 Representation of a MMX chain based of on platinum, dithiocarboxylate and iodine, $[Pt_2(RCS_2)_4I]_n$ (R = alkyl group).

More recently, studies on $[Pt_2(RCS_2)_4I]_n$ (R = alkyl group) chains have shown the high potential of these materials as nano and molecular wires. In a first work, the electrical behaviour of single crystals of $[Pt_2(n\text{-pentylCS}_2)_4I]_n$ has been studied. The data have shown conductivity values in the range 0.3–1.4 $S\cdot cm^{-1}$ at room temperature.[24] Variable temperature measurements carried out on several crystals following this treatment: (i) heated from 300 to 400 K, (ii) cooled to 100 K, and finally (iii) heating from 100 to 400 K; showed that the first heating scan displayed a metallic behaviour from 300 to 400 K with a transition (RT-HT) that appears as a drop in the resistivity with a negative peak in the derivative at 330 K (Figure 11a). Above this RT-HT transition, the sample recovers its metallic character and reaches a resistivity of *ca.* 7 $\Omega\cdot cm$ at 400 K. The transition at 330 K has also been observed in heat capacity measurements, although at a temperature of 324 K, and has been attributed to a moderate disorder of the alkyl chains.[25] When the temperature is decreased from 400 to 100 K, the resistivity smoothly increases to reach a maximum at *ca.* 390 K (Figure 11a) followed by metallic behaviour from 390 K down to *ca.* 290 K, where it shows a rounded minimum, similar to that shown by the non-heated crystals, although at a higher temperature.

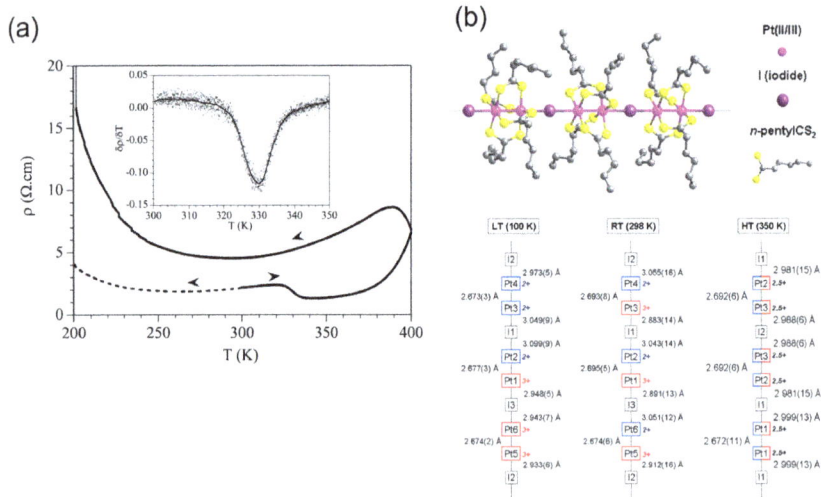

Figure 11   (a) Thermal variation of the electrical resistivity of $[Pt_2(n\text{-pentylCS}_2)_4I]_n$. The dashed line displays the behaviour of a non-heated sample. The inset displays the derivative of the resistivity as a function of the temperature around the RT-HT transition. (b) View of the polymeric $[Pt_2(n\text{-pentylCS}_2)_4I]_n$ chain (top) and bond distances distribution along the chain at 100, 298 and 350 K (bottom).

This behaviour confirms the irreversibility of the RT-HT transition. A detailed X-ray diffraction analyses, carried out at different temperatures, confirms the existence of three different phases for $[Pt_2(n\text{-pentylCS}_2)_4I]_n$ (Figure 11b). The crystal structures collected at 100 (LT), 298 (RT) and 350 (HT) K confirm that in all cases linear 1D chains generated by repetition of the MMX entities. Each pair of platinum atoms is bridged by four n-pentyl-CS_2 ligands, and the pairs are interconnected by means of linear iodine bridges. The alkyl chain surrounds the Pt–Pt–I–chain precluding the presence of S$\cdots$S contacts. The three-dimensional cohesiveness of the crystal structure is therefore ensured only by means of weak interchain Van der Waals interactions among the alkyl chains. Finally, Density Functional Theory (DFT) calculations allowed the characterization of three different valence-ordering states.

4.2 In Solution

The question about what are the species formed in solution by disolving MMX crystals and what are the experimental factors affecting the MMX chains formation from these solutions, are essential aspects to be solved in order to understand the chemistry of these molecules and to develop methods to process these materials. The fact that thin films of MMX are almost transparent in thickness, < 50 nm, make them extremely appealing for molecular electronics and optoelectronics.

The outstanding ability of the $[Pt_2(n\text{BuCS}_2)_4I]_n$ ($n\text{Bu} = $ n-butyl) chains to reversibly self-organize from solution is an uncommon behaviour for other metal-organic polymers and raises the questions of which species are present in solution, and which factors have an effect on their subsequent reassembly.[26] In order to address these important factors, the spectroscopic features of $[Pt_2(n\text{BuCS}_2)_4I]_n$ dissolved in CH_2Cl_2 were analysed and compared with those of the $[Pt_2(n\text{BuCS}_2)_4]$ and $[Pt_2(n\text{BuCS}_2)_4I_2]$ precursors. The data showed that, at room temperature, $[Pt_2(n\text{BuCS}_2)_4I]_n$ dissociate in solution into an equimolar mixture of $[Pt_2(n\text{BuCS}_2)_4]$ and $[Pt_2(n\text{BuCS}_2)_4I_2]$. The UV-vis spectrum of a 1 mM solution of $[Pt_2(n\text{BuCS}_2)_4I]_n$ can be understood as the result of overlapping the UV-vis spectra separately measured from 0.5 mM CH_2Cl_2 solutions of precursors $[Pt_2(n\text{BuCS}_2)_4]$ and $[Pt_2(n\text{BuCS}_2)_4I_2]$. Consistently, the ^1H NMR spectrum of $[Pt_2(n\text{BuCS}_2)_4I]_n$ in CD_2Cl_2 shows an overlapping of the spectra of species $[Pt_2(n\text{BuCS}_2)_4]$ and $[Pt_2(n\text{BuCS}_2)_4I_2]$. The spectroscopic data indicate an asymmetric rupture of $[Pt_2(n\text{BuCS}_2)_4I]_n$, which formed two different diamagnetic dimetallic compounds containing two Pt(II) centres (in the case of $[Pt_2(n\text{BuCS}_2)_4]$) or two Pt(III) centres (for $[Pt_2(n\text{BuCS}_2)_4I_2]$) (Figure 12).

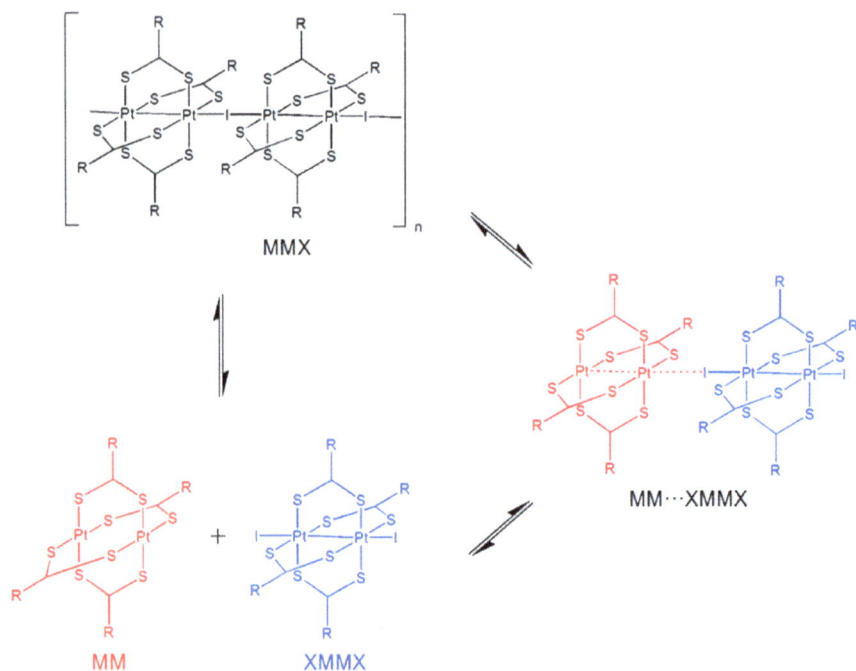

Figure 12   Representation of the dissociation and reassembly processes in solution of crystals of $[Pt_2(RCS_2)_4I]_n$ (R = alkyl group) chains.

Concerning the (re)assembly of $[Pt_2(nBuCS_2)_4]$ and $[Pt_2(nBuCS_2)_4I_2]$ in solution, above we have mentioned the ability of the $[Pt_2(nBuCS_2)_4]$ precursor to associate *via* reversible weak $d^8\cdots d^8$ interactions. This can be observed by the appearance at low temperature of an adsorption band at 600–700 nm in its UV-vis spectrum. Similar behaviour was observed for solutions of $[Pt_2(nBuCS_2)_4I]_n$. Additionally, a new band at 820 nm appeared in the spectrum of $[Pt_2(nBuCS_2)_4I]_n$ in CH_2Cl_2 at –50 °C. Considering that this feature was not observed for the solutions of pure precursors $[Pt_2(nBuCS_2)_4]$ or $[Pt_2(nBuCS_2)_4I_2]$, this new band indicates the assembly of molecules $[Pt_2(nBuCS_2)_4]$ and $[Pt_2(nBuCS_2)_4I_2]$ generating oligomers of the $[Pt_2(nBuCS_2)_4I]_n$ chains. The ratio between the $[Pt_2(nBuCS_2)_4]\cdots$ $[Pt_2(nBuCS_2)_4]$ and the $[Pt_2(nBuCS_2)_4]\cdots[Pt_2(nBuCS_2)_4I_2]$ association was affected by the overall concentration of dissolved $[Pt_2(nBuCS_2)_4I]_n$, being the $[Pt_2(nBuCS_2)_4]\cdots[Pt_2(nBuCS_2)_4]$ assembly less favoured at lower concentrations. The low temperature experiments offer valuable insights that help to understand the process of MMX polymer formation from solution.

4.3 From Solution to Surface

Fibres of $[Ru_2Br(\mu\text{-}O_2CEt)_4]_n$ chains were isolated on different surfaces by casting deposition of water-sodium dodecyl sulphatesolutions.[27] It seems feasible that the evaporation of highly diluted solutions of $[Ru_2Br(\mu\text{-}O_2CEt)_4]_n$ on the surfaces led to fibre formation. AFM topography images allowed the characterization of homogeneous fibres with lengths from 0.5 to 5 mm and typical heights of *ca.* 0.7 nm, corresponding to the isolation on mica of individual polymer chains. The analysis of solutions of $[Ru_2Br(\mu\text{-}O_2CR)_4]_n$ chains suggested formation of dicationic subunits $RuRu^+$ and Br^-. A more detailed study on $[Ru_2Br(\mu\text{-}O_2CEt)_4]_n$showed a mechanochemistry based procedure to isolate individual polymer chains on surfaces with subnanometer diameter and over microns length. They studied the dynamics process that take place after sonicated solutions of $[Ru_2Br(\mu\text{-}O_2CEt)_4]_n$.[28] They demonstrate that ultrasound induces scission of the weaker coordination bonds giving rise to reactive species that self-organize in solution allowing a rich variety of structures. The recognition of these activated building blocks leads to isolate very long individual chains of the MMX polymer when adsorbed on surfaces as demonstrated by the AFM images. It seems that the linear observed structures reproduce those present in the solution. This work suggested a way to form well-organized entities of $[Ru_2Br(\mu\text{-}O_2CR)_4]_n$chains on surfaces.

Two different nanostructures were characterized on mica from the controlled deposition of sonicated diluted THF solutions of $[Pt_2(n\text{-}pentylCS_2)_4I]_n$. Using this simple procedure, high density coverage of micron-length fibres of $[Pt_2(n\text{-}pentylCS_2)_4I]_n$ with typical height of *ca.* 1.5 to 2.5 nm were isolated. According to the X-ray data (1.5 nm being the expected height of one MMX chain), these fibres correspond to a few MMX chains. Additionally, upon increasing the concentration of the initial $[Pt_2(n\text{-}pentylCS_2)_4I]_n$ solution, nanocrystals were formed on mica. AFM topographic image of several nanocrystals growing from a central nucleus showed a length of 1–5 μm and a typical height of 2–4 nm (Figure 9).

While formation of structures from solution seems to be a simple and powerful method, at that point the understanding of the parameters controlling both the molecular association in solution and the deposition were unclear.

The rather uncommon feature, dissolution and repolymerization conserving its structural integrity, showed for the $[Pt_2(nBuCS_2)_4I]_n$ chains enables its processability and the formation of organized structures on surfaces. Thus, micromolding in capillaries (MIMIC)[29] and lithographically controlled wetting (LCW)[30] were successfully applied to form a variety of nanostructures (Figure 13).

Figure 14 shows an optical micrograph of the interdigitated electrodes of a FET printed by LCW and parallel μ-wires on pre-fabricated gold electrodes that were printed by MIMIC. The electrical characterization performed by measuring the current flowing in the μ-wires as a function of bias voltage, ranging from +50 to –50 V reveals a near Ohmic behaviour. It is important to note that the time stability of the wires is longer than 6 months, in air at room temperature and in high humidity. As an example of direct application of $[Pt_2(nBuCS_2)_4I]_n$ nanostructures, the use of this material as electrodes in organic field-effect transistor (OFET) devices was tested (Figure 15)[31] Thus,

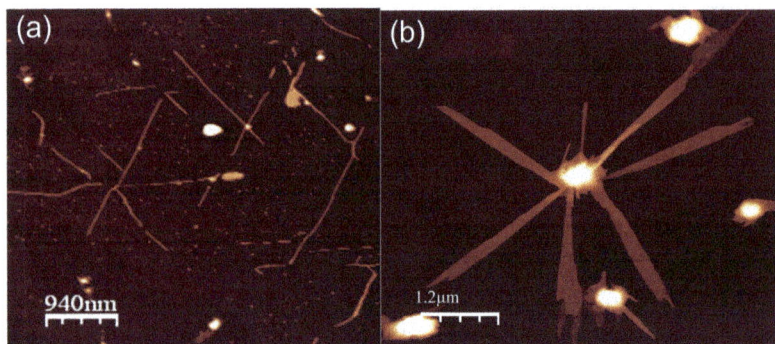

Figure 13   Topographic AFM images of nanostructures of $[Pt_2(n\text{-pentyl}CS_2)_4I]_n$ showing over-micron length fibres of *ca.* 1.5 to 2.5 nm of height (a) and flat nanocrystals of 2–4 nm height (b).

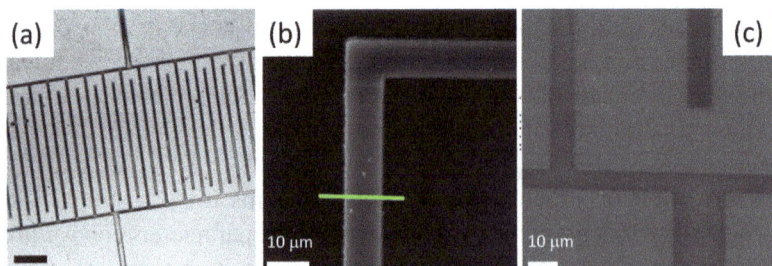

Figure 14   Some microfabricated, by unconventional wet lithography, structures of $[Pt_2(nBuCS_2)_4I]_n$. a) Optical micrograph (scale bar = 100 mm) of interdigitated comb-like electrodes printed on silicon oxide by LCW, b) AFM image and c) SEM image of $[Pt_2(nBuCS_2)_4I]_n$ pattern on silicon oxide.

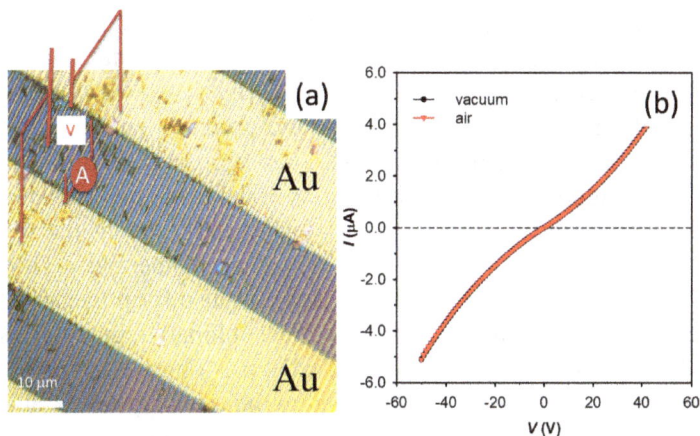

Figure 15 a) Optical image of parallel μ-wires of $[Pt_2(nBuCS_2)_4I]_n$ printed on silicon oxide by unconventional wet lithography onto Au electrodes. b) Current *vs.* voltage characteristics of the $[Pt_2(nBuCS_2)_4I]_n$ wires shown in (a).

structures produced by MIMIC technique on SiO_2/Si from a $[Pt_2(nBuCS_2)_4I]_n$ solution demonstrated that this material can be efficiently used as electrodes for organic field-effect transistor. The OFET devices showed a field-effect performance with charge mobility, μ_{sat}, up to 5.2×10^{-3} $cm^2V^{-1}s^{-1}$, with on/off ratios around 10^6 and an excellent stability under ambient conditions (after six months from fabrication the device remains almost unaltered).

4.4 From solid to gas phase to surface assembly

Nanostructures obtained from $[Pt_2(n\text{-}pentylCS_2)_4I]_n$ solutions were not suitable to measure the electrical resistance using a macroscopic gold electrode evaporated with a conventional mask technique.[32] Apparently, the nanostructures formed by drop-casting from the polymer solution, contain some amount of solvent retained on the surface in the adsorption process together with the MMX nanostructures. Under a vacuum during the gold sublimation process, the adsorbed solvent rapidly evaporates generating fractures across the MMX nanostructures and avoiding a proper characterization. Therefore, a less conventional approach based on direct sublimation from synthetized monocrystals of MMX was explored. This method has been successfully applied to the adsorption of $[Pt_2(n\text{-}butylCS_2)_4I]_n$ on several surfaces.[33] The procedure consists of (*i*) sublimation of $[Pt_2(n\text{-}butylCS_2)_4I]_n$ crystals in a high vacuum chamber (10^{-6} mbar), and then (*ii*) the landing of the sublimated materialon a substrate ata controlled temperature. The analysis by AFM of the surfaces reveals formation of 1D structures, with lengths of several microns and typical heights around 5 to 10 nm when the parameters were properly adjusted. The sublimation method produces straighter fibres with more homogeneous width distributions than drop casting procedures. Additionally, it was reported that the dimensions of the structures obtained by sublimation can be modulated by controlling parameters such as the nature of the surface, the deposition time and the annealing temperature. Thus, it has not only been possible to isolate nanofibres, but also form nanocrystals that have very few defects. The mechanism of the sublimation process is still not well established, however, in analogy to what it is observed in solution, it seems likely that the thermal energy applied to the monocrystals is able to induce the breakage of the weaker coordinative bonds leading to the formation of small entities which are volatile under the experimental conditions to fly and land on the surface. Subsequently, these small entities will diffuse and self-assemble to form the MMX chains. This is the description of a reversible process consequence of the reversibility feature of a typical coordinative bond which is the basic feature of a MMX structure.

More recently, the high potential of MMX chains towards their use as molecular wires has been proved. Thus, nanoribbons of $[Pt_2(dta)_4I]_n$ isolated on mica [34] and SiO_2[35] by direct sublimation from crystals have shown outstanding electrical properties.

The conductive nanoribbons were formed on mica by direct sublimation from crystals of $[Pt_2(dta)_4I]_n$ under a high vacuum (Figure 16). However, while these nanostructures showed high conductivity the non-linear IV features suggested a significant content of defect in these nanoribbons, therefore limiting their conductivity (Figure 17).[34]

Recently, following a similar experimental procedure to that previously used on mica, $[Pt_2(dta)_4I]_n$ nanoribbons have been isolated on SiO_2. Those nanoribbons are larger but more important they have shown high order, defect-free for distances below *ca.* 300 nm.[35] The conductivity for these nanostructures is 10^4 S/m, three orders of magnitude higher than that of our macroscopic crystals. This magnitude is preserved for distances as large as 300 nm. Above this length, the presence of structural defects ($\sim$ 0.5 %) gives rise to an inter-fibre mediated charge transport similar to that of macroscopic crystals. Additional experiments show the first direct experimental evidence of the gapless electronic structure theoretically predicted [36] for $[Pt_2(dta)_4I]_n$ chains.

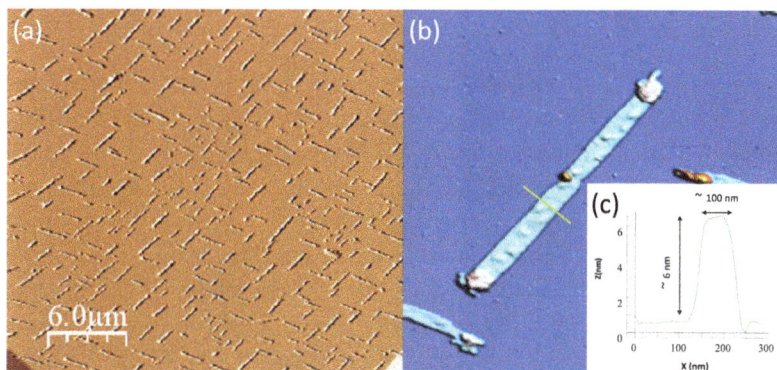

Figure 16 AFM topography image of over micron length nanoribbons formed on mica by direct sublimation of $[Pt_2(dta)_4I]_n$ monocrystals (a). A zoomed nanoribbon (b) and its height profile (c)

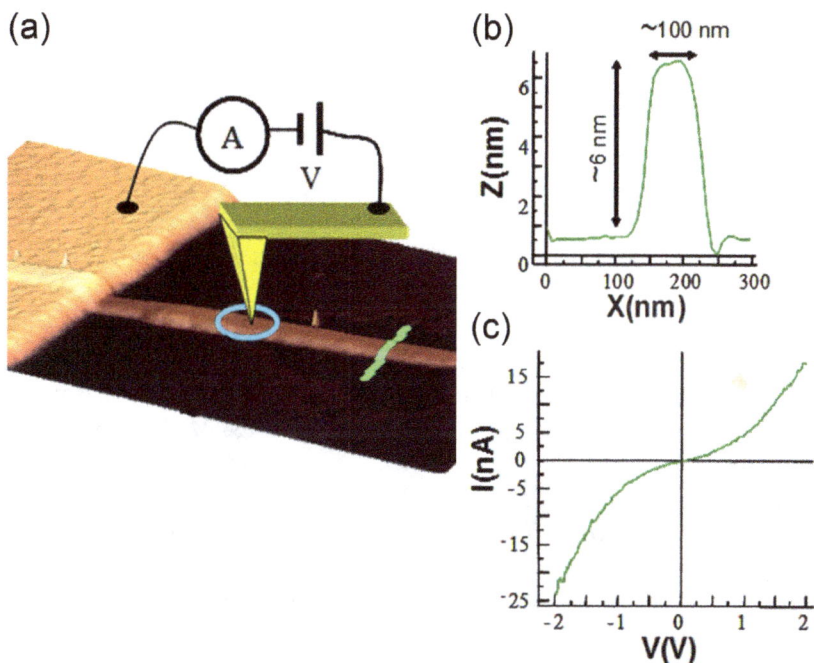

Figure 17 (a) AFM topography showing a MMX nanoribbon adsorbed on mica and connected to a gold electrode. The nanoribbon is partially covered with gold macro-electrode. (b) AFM height profile taken along the green line drawn in (a). (c) Current *vs.* voltage features taken by contacting the nanoribbon with a gold AFM tip located at 100 nm from the gold macro-electrode.

5 Conclusions

According the bibliographic information presented in this review metal$\cdots$metal interactions, Au(I)$\cdots$Au(I) and Pt(II)$\cdots$Pt(II), are the source of spontaneous self-assembly processes. In solution, reversible aggregation of tetra-, hexa- and octa-metallic supramolecules has been observed at low temperatures. These aggregations have different consequences in the solid state structures. For Au(I) containing structures, Au$\cdots$Au interactions trigger oligomerization to generate oligomeric/polymeric $[Au_2L_2]_n$ (L = dithiocarboxylato) structures by means of the μ-kS:kS' bridging mode. In contrast, Pt(II) compounds show linear arrangements hold together exclusively through Pt$\cdots$Pt interactions. Crystals of $[Pt_2L_4]$ show electrical behaviour characteristic of semiconductors. In addition, supramolecular Pt$\cdots$Pt assembly triggers formation of 1D nanostructures on surfaces. Furthermore, partial oxidation of $[Pt_2L_4]$ with iodine generates supramolecular 1D assemblies of formula $[Pt_2L_4I]_n$ based on week metal-ligand interactions. Assembly and disassembly of $[Pt_2L_4I]_n$ conductive structures, even at the nanoscale, allows to envisage future technological applications in molecular electronics.

References

[1] E. R. Kay, D. A. Leigh and F. Zerbetto, *Angew. Chem. Int. Ed.*, **46**, 72–191 (2007).

[2] J. M. Lehn, *Proc. Nat. Acad. Sci. USA*, **99**, 4763–4768 (2002).

[3] a) J. K. Bera and K. R. Dunbar, *Angew. Chem. Int. Ed.*, **41**, 4453–4458 (2002); b) F. A. Cotton, C. A. Murillo and R. A. Walton, *Multiple Bonds Between Metal Atoms*, *Springer Science and Business Media Inc.*, New York (2005).

[4] S. Roth, *One-dimensional Metals*, *VCH*, New York (1995).

[5] M. Williams, *Adv. Inorg. Chem. Radiochem.*, **26**, 235–268 (1983).

[6] a) R. Mas-Balleste, J. Gomez-Herrero and F. Zamora, *Chem. Soc. Rev.*, **39**, 4220–4233 (2010); b) J. Gomez-Herrero and F. Zamora, *Adv. Mater.*, **23**, 5311–5317 (2011).

[7] F. A. Cotton, C. Lin and C. A. Murillo, *Acc. Chem. Res.*, **34**, 759–771 (2001).

[8] F. A. Cotton and R. A. Walton, *Multiple Bonds Between Metal Atoms*, 2nd ed., Clarendon, Oxford (1993).

[9] a) R. Mas-Balleste, R. Gonzalez-Prieto, A. Guijarro, M. A. Fernandez-Vindel and F. Zamora, *Dalton Trans.*, 7341–7343 (2009); b) A. Guijarro,

O. Castillo, A. Calzolari, P. J. S. Miguel, C. J. Gomez-Garcia, R. di Felice and F. Zamora, *Inorg. Chem.*, **47**, 9736–9738 (2008); c) A. Kobayashi, T. Kojima, R. Ikeda and H. Kitagawa, *Inorg. Chem.*, **45**, 322–327 (2006).

[10] a) H. Schmidbaur and A. Schier, *Chem. Soc. Rev.*, **37**, 1931–1951 (2008); b) P. Pyykko, *Chem. Soc. Rev.*, **37**, 1967–1997 (2008).

[11] a) A. Vogler and H. Kunkely, *Chem. Phys. Lett.*, **150**, 135–137 (1988); b) P. Pyykko and Y. F. Zhao, *Angew. Chem. Int. Ed.*, **30,** 604–605 (1991); c) Y. Jiang, S. Alvarez and R. Hoffmann, *Inorg. Chem.*, **24**, 749–757 (1985); d) P. K. Mehrotra and R. Hoffmann, *Inorg. Chem.*, **17**, 2187–2189 (1978); e) A. Dedieu and R. Hoffmann, *J. Am. Chem. Soc.*, **100**, 2074–2079 (1978); f) K. M. Merz and R. Hoffmann, *Inorg. Chem.*, **27**, 2120–2127 (1988); g) M. A. Carvajal, S. Alvarez and J. J. Novoa, *Chem. Eur. J.*, **10**, 2117–2132 (2004).

[12] a) V. W. W. Yam and E. C. C. Cheng, *Top. Curr. Chem.*, **281**, 269–309; b) R. J. Puddephatt, Coord. *Chem. Rev.* 2001, 216, 313–332 (2007).

[13] a) V. W. W. Yam and E. C. C. Cheng, *Chem. Soc. Rev.*, **37**, 1806– 1813 (2008); b) H. Schmidbaur and A. Schier, *Chem. Soc. Rev.*, **41**, 370–412 (2012).

[14] a) M. L. Gallego, A. Guijarro, O. Castillo, T. Parella, R. Mas-Balleste and F. Zamora, *Cryst. Eng. Commun.*, **12**, 2332–2334 (2010); b) M. R. Azani, O. Castillo, M. L. Gallego, T. Parella, G. Aullon, O. Crespo, A. Laguna, S. Alvarez, R. Mas-Balleste and F. Zamora, *Chem. Eur. J.*, **18**, 9965–9976 (2012).

[15] a) D. D. Heinrich, J. C. Wang and J. P. J. Fackler, Acta Crystall. C, **46**, 1444 (1990); b) M.A. Mansour, W.B. Connick, R. J. Lachicotte, H. J. Gysling and R. Eisenberg, *J. Am. Chem. Soc.*, **120**, 1329 (1998); c) M. A. Mansour, W. B. Connick, R. J. Lachicotte, H. J. Gysling and R. Eisenberg, *J. Am. Chem. Soc.*, **120**, 1329 (1998); d) R. Hesse and P. Jennische, Acta Chem. Scand., **26**, 3855 (1972); e) S. Y. Ho and E. R. T. Tiekink, *Z. Kristall. New Cryst. Struct.*, **217**, 589 (2002).

[16] a) C. Bellitto, M. Bonamico, G. Dessy, V. Fares and A. Flamini, *J. Chem. Soc. Dalton Trans.*, 35–40 (1987); b) C. Bellitto, G. Dessy, V. Fares and A. Flamini, *J. Che. Soc. Chem. Commun.*, 409–411 (1981).

[17] A. P. Paz, L. A. Espinosa Leal, M. R. Azani, A. Guijarro, P. J. S. Miguel, G. Givaja, O. Castillo, R. Mas-Balleste, F. Zamora and A. Rubio, *Chem. Eur. J.*, **18**, 13787–13799 (2012).

[18] T. Kawamura, T. Ogawa, T. Yamabe, H. Masuda and T. Taga, *Inorg. Chem.*, **26**, 3547–3550 (1987).

[19] M. Yamashita and H. e. Okamoto, *Material Designs and New Physical Properties in MX- and MMX-Chain Compounds*, Springer-Verlag Wien (2013).

[20] a) M. C. Barral, R. Gonzalez-Prieto, R. Jimenez-Aparicio, J. L. Priego, M. R. Torres and F. A. Urbanos, *Eur. J. Inorg. Chem.*, 2339–2347 (2003); b) M. C. Barral, R. Gonzalez-Prieto, R. Jimenez-Aparicio, J. L. Priego, M. R. Torres and F. A. Urbanos, *Eur. J. Inorg. Chem.*, 4491–4501 (2004); c) M. C. Barral, R. Jimenez-Aparicio, D. Perez-Quintanilla, J. L. Priego, E. C. Royer, M. R. Torres and F. A. Urbanos, *Inorg. Chem.*, **39**, 65–70 (2000).

[21] a) H. Kitagawa, N. Onodera, T. Sonoyama, M. Yamamoto, T. Fukawa, T. Mitani, M. Seto and Y. Maeda, *J. Am. Chem. Soc.*, **121**, 10068–10080 (1999); b) M. Mitsumi, T. Murase, H. Kishida, T. Yoshinari, Y. Ozawa, K. Toriumi, T. Sonoyama, H. Kitagawa and T. Mitani, *J. Am. Chem. Soc.*, **123**, 11179–11192 (2001).

[22] M. Mitsumi, K. Kitamura, A. Morinaga, Y. Ozawa, M. Kobayashi, K. Toriumi, Y. Iso, H. Kitagawa and T. Mitani, *Angew. Chem. Int. Ed.*, **41**, 2767–2771 (2002).

[23] Y. Wakabayashi, A. Kobayashi, H. Sawa, H. Ohsumi, N. Ikeda and H. Kitagawa, *Journal of the American Chemical Society*, **128**, 6676–6682 (2006).

[24] A. Guijarro, O. Castillo, L. Welte, A. Calzolari, P. J. S. Miguel, C. J. Gomez-Garcia, D. Olea, R. di Felice, J. Gomez-Herrero and F. Zamora, *Adv. Funct. Mater.*, **20**, 1451–1457 (2010).

[25] K. Saito, S. Ikeuchi, Y. Nakazawa, A. Sato, M. Mitsumi, T. Yamashita, K. Toriumi and M. Sorai, *J. Phys. Chem. B*, **109**, 2956–2961 (2005).

[26] D. Gentili, G. Givaja, R. Mas-Balleste, M. R. Azani, A. Shehu, F. Leonardi, E. Mateo-Marti, P. Greco, F. Zamora and M. Cavallini, *Chem. Sci.*, **3**, 2047–2051 (2012).

[27] D. Olea, R. Gonzalez-Prieto, J. L. Priego, M. C. Barral, P. J. de Pablo, M. R. Torres, J. Gomez-Herrero, R. Jimenez-Aparicio and F. Zamora, *Chem. Commun.*, 1591–1593 (2007).

[28] L. Welte, R. González-Prieto, D. Olea, M. Rosario Torres, J. L. Priego, R. Jiménez-Aparicio, J. Gómez-Herrero and F. Zamora, *ACS Nano*, **2**, 2051–2056 (2008).

[29] a) E. Kim, Y. N. Xia and G. M. Whitesides, *Nature*, **376**, 581–584 (1995); b) M. Cavallini, C. Albonetti and F. Biscarini, *Adv. Mater.*, **21**, 1043–1053 (2009).

[30] a) M. Cavallini and F. Biscarini, Nano Letters, **3**, 1269–1271 (2003);b) M. Cavallini, D. Gentili, P. Greco, F. Valle and F. Biscarini, *Nat. Prot.*, **7**, 1668–1676 (2012).

[31] H. B. Akkerman, P.W. M. Blom, D. M. de Leeuw and B. de Boer, *Nature*, **441**, 69–72 (2006).

[32] P. J. de Pablo, M. T. Martinez, J. Colchero, J. Gomez-Herrero, W. K. Maser, A. M. Benito, E. Munoz and A. M. Baro, *Adv. Mater.*, **12**, 573–576 (2000).

[33] L.Welte, U. García-Couceiro, O. Castillo, D. Olea, C. Polop, A. Guijarro, A. Luque, J. M. Gómez-Rodríguez, J. Gómez-Herrero and F. Zamora, *Adv. Mater.*, **21**, 2025–2028 (2009).

[34] L. Welte, A. Calzolari, R. di Felice, F. Zamora and J. Gómez-Herrero, *Nat. Nanotech.*, **5**, 110–115 (2010).

[35] C. Hermosa, J. V. Álvarez, M. R. Azani, C. J. Gómez-García, M. Fritz, J. M. Soler, J. Gómez-Herrero, C. Gómez-Navarro and F. Zamora, *Nat. Commun.*, **4**, 1709 (2013).

[36] A. Calzolari, S. S. Alexandre, F. Zamora and R. Di Felice, *J. Am. Chem. Soc.*, **130**, 5552–5562 (2008).

Biographies

Rubén Mas-Ballesté was born in Barcelona (Catalonia, Spain) in October of 1975. In 2004 he got his Ph.D. under the supervision of Prof. Pilar González-Duarte and Prof. Agustí Lledós at Universitat Autónoma de Barcelona. From June 2004 to November 2007 he was working in the Lawrence Que's group at the University of Minnesota (USA) as a postdoctoral associate. At the present he is an associate professor at the Universidad Autónoma de Madrid (Spain) where he was appointed under the "Ramón y Cajal" program in 2008. His research is focused on activation/formation of small molecules, synthesis and reactivity of 2D organic polymers and inorganic materials for nanotechnology.

Félix Zamora was born in 1967 in Cuenca (Spain). In 1994 he obtained the PhD in Inorganic Chemistry at Universidad Autónoma de Madrid. He moved to University of Dortmund (Germany) to work with Professor B. Lippert. He is currently "Profesor Titular" at the Inorg. Chem. Department at the Universidad Autónoma de Madrid. From 2004 he focused on new nanomaterials based on inorganic systems such as coordination polymers.

Integrating DNA with Functional Nanomaterials

Shalom J. Wind[1], Erika Penzo[1], Matteo Palma[1], Risheng Wang[1],
Teresa Fazio[1], Danny Porath[2], Dvir Rotem[2],
Gideon Livshits[2] and Avigail Stern[2]

[1]*Department of Applied Physics and Applied Mathematics Columbia University,
New York, NY, USA* [2]*Department of Chemistry Hebrew University of Jerusalem,
Jerusalem, Israel*

Received 10 May 2013; Accepted 22 May 2013; Publication 6 June 2013

1 Introduction

DNA may be the most versatile molecule discovered to date. Beyond its well-known central role in genetics, DNA has the potential to be a remarkably useful technological material. It has been demonstrated as a scaffold for the assembly of organic and inorganic nanomaterials [1]; a vehicle for drug delivery [2]; a medium for computation [3]; and a possible wire for transporting electrical signals [4]. A key factor in exploiting DNA in these ways is the ability to integrate DNA with other materials. In this paper, we review two approaches to forming DNA complexes with functional nanomaterials: (1) linking DNA with single-wall carbon nanotubes (SWCNTs), which can then be used as nanoscale electrical contacts for probing electron transport in DNA; and (2) directed nanoassembly of Au nanoparticles using DNA/PNA (peptide nucleic acid) hybrid scaffolds.

2 DNA-SWCNT Junctions

This work is motivated by the original work of Guo et al. [5, 6], which demonstrated the efficacy of SWNT electrodes for the study of charge transport through individual molecules. SWNTs are nearly ideal for this purpose. They are outstanding one-dimensional conductors, they can be linked to organic

Journal of Self-Assembly and Molecular Electronics, Vol. 1, 177–194.
doi: 10.13052/jsame2245-4551.122

molecules through straightforward carbon-carbon chemistries, and they are essentially the same size (diameter) as individual molecules, ensuring that only a single molecule is being probed in each experiment. This platform was first applied to the study of charge transport in DNA by the Nuckolls group at Columbia, where they demonstrated efficient transport through well-matched dsDNA strands connected to SWCNTs via an amine linkage, supporting the contention that dsDNA contacted in this way maintains its native conformation [7]. The single-molecule devices in this and the previous work were fabricated by a process in which a nanoscalegap in a SWCNT is formed by "cutting" the SWCNT through a lithographically-defined stencil using an O_2 plasma. This approach has been quite successful, in terms of demonstrating electron transport in DNA and other molecules, however, it is extremely inefficient; only ~3% or fewer of cut nanotubes result in reconnection with the DNA. The primary reason for this is the difficulty in precisely matching the size of the opening to the length of the dsDNA molecule.

In order to overcome this difficulty, we have taken a new approach in which the most challenging aspect of the process, namely, formation of the connection between the DNA molecule and the SWCNT electrodes, would be achieved in solution by chemical means. Once these hybrid structures are formed, they could be placed on a surface for electrical measurement, using either a shadow mask electrode in conjunction with a conductive AFM tip (a technique developed by the Porath lab) or using pre-patterned electrodes on the surface. The extended length of the SWCNT-dsDNA hybrids renders them far easier to contact than individual DNA molecules.

SWCNT-dsDNA hybrid complexes (Fig. 1) were created by reacting a water solution of SWCNT segments with amine functionalized dsDNA (26 base pairs mixed sequence, described below). Two reaction schemes were developed, one consisting of one step, the other consisting of two steps. The second scheme results in a higher yield of the desired CNT-dsDNA-CNT structure (Fig. 1).

The starting material consisted of short SWCNT segments wrapped in single stranded DNA [DNA sequence: $(GT)_{20}$] and dissolved in deionized water (concentration ~40 μg/ml), obtained from M. Zheng at NIST. The solution is

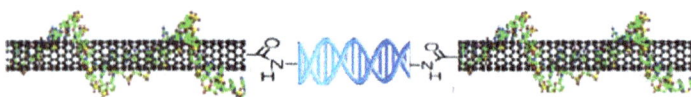

Figure 1   Schematic of end-connected SWCNT-dsDNA Hybrid Structure.

the result of a purification procedure based on size-exclusion chromatography (SEC) which sorts the CNT segments into fractions of uniform length [8]. The SWCNT segment length distribution was quantified by tapping mode AFM imaging (Fig. 2a) and software analysis (ImageJ, Fig. 2b). AFM samples were obtained by depositing 10 μl of CNT solution diluted 1:20 in deionized water, on a silicon dioxide substrate treated with oxygen plasma. The solution was dried in air and the samples were then washed by dipping them for 10 seconds in a solution of 50% DI water, 50% ethanol, then immersing them in a solution of 10% DI water, 90% ethanol for 50 minutes and finally letting them dry in air. The average length and standard deviation of the CNT segments in the starting solution were found to be 148 $\pm$ 93 nm. Using this average length we estimated the molar concentration of CNT segments to be about 70 nM.

Two different schemes were followed to create the SWCNT-dsDNA hybrids. In the first scheme, a one-step reaction, the starting SWCNT solution was activated by mixing it, 1:1 by volume, with a solution consisting of 0.2 M MES buffer (pH 6), 4 mM EDC and 10 mM sulfo-NHS. The SWCNTs sat in this solution for 30 min. at room temperature, during which the EDC and sulfo-NHS form an intermediate compound with the carboxyl groups on the CNT ends. Following activation, 2 μl of 0.5 μM amine-functionalized dsDNA (amine-26bp, described above) was added. The intermediate compound reacts with the amine groups on the DNA strands resulting in a covalent bond between

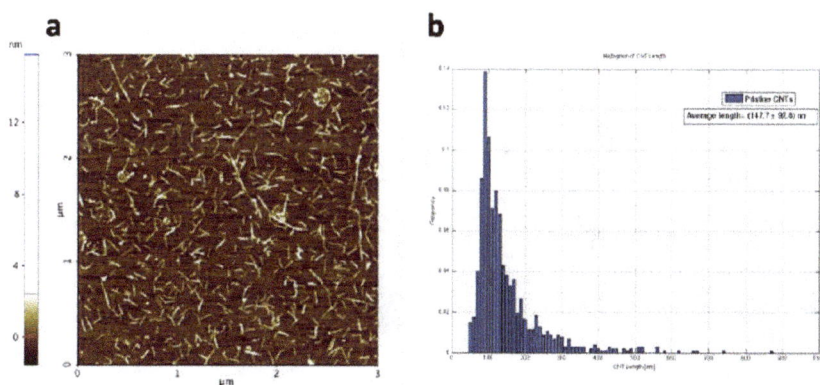

Figure 2 a) Tapping mode AFM image of the pristine DNA-wrapped SWCNT segments deionized water solution. (b) Histogram of the nanotube segments length distribution of the starting solution. The histogram was built by measuring the SWCNT length with ImageJ software. 1024 nanotube segments were measured, from different AFM images of different substrates on which the same solution was deposited. The average length and standard deviation of the SWCNT segments in the starting solution were found to be 148 $\pm$ 93 nm

the dsDNA and the SWCNT segments. This method for DNA attachment to acid-oxidized carbon nanotubes ends was previously reported by Weizmann et al. [10]. The concentration of amine-26bp during the reaction was 5 nM, corresponding to about one seventh of the nanotube concentration. The difference in concentration increases the probability that each DNA strand will react with two nanotube segments, one on each side, yielding the desired SWCNT-dsDNA hybrid structure. The mixture was left to react overnight at room temperature. Any unreacted DNA strands were removed by centrifugation in Millipore Amicon 100K tubes. During purification the buffer was exchanged to DPBS 1X.

A histogram showing the length distribution of the resulting SWCNT-dsDNA hybrid structures (Fig. 3) was obtained by tapping mode AFM imaging and software analysis as explained above for the starting SWCNT solution. The average length and standard deviation of the CNT-dsDNA synthesized by this one step reaction were 325 ± 283 nm.

A second, two-step reaction scheme was also investigated. In this scheme, 20 μl of the starting SWCNT solution was mixed, 1:1 by volume, with a solution consisting of 0.2 M MES buffer, 4 mM EDC and 10 mM sulfo-NHS. This solution was let to activate for 30 min at room temperature before adding 20 μl of 0.5 μM amine-functionalized dsDNA (amine-26bp). The resulting concentration of amine-26bp was 167 nM, making it sufficiently likely that

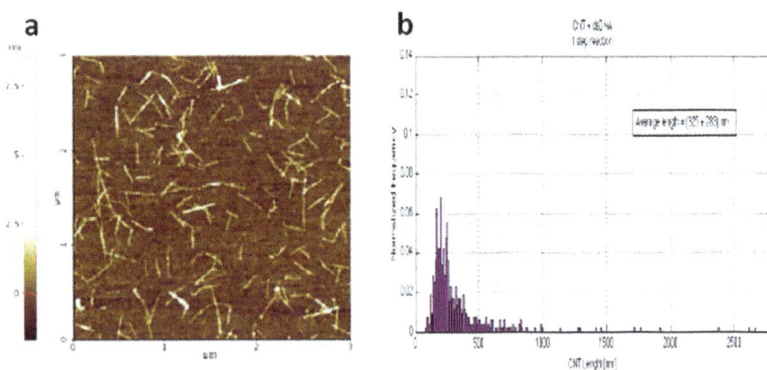

Figure 3   (a) Tapping mode AFM image of the SWCNT-dsDNA solution generated by the one-step reaction. (b) Histogram of the length distribution of the CNT-dsDNA hybrid structures obtained with the one-step reaction. The histogram was built by measuring the SWCNT length with ImageJ software. 525 nanotube segments were measured, from different AFM images of different substrates on which the same solution was deposited. The average length and standard deviation of the CNT segments in the starting solution were found to be 325 ± 283 nm.

both ends of all SWCNT segments would be saturated. The mixture was left to react overnight at room temperature. Any unreacted DNA strands were removed by centrifugation in Millipore Amicon 100K tubes (the residual concentration of dsDNA after purification was estimated to be less than 0.5 nM). During purification the buffer was exchanged to DPBS 1X. 20 μl of the starting SWCNT solution was mixed, 1:1 by volume, to a solution consisting of 0.2 M MES buffer, 4 mM EDC and 10 mM sulfo-NHS. This solution was left to activate for 30 min. at room temperature before adding it to the same volume of the purified SWCNT-dsDNA. This second reaction resulted in the attachment of the activated SWCNT segments to the ones previously bound to the dsDNA.

The size distribution of the resulting CNT-dsDNA hybrid structures is shown in Fig. 4. The average length and standard deviation of the CNT-dsDNA synthesized by this two steps reaction are 248 $\pm$ 113 nm.

Both the one-step and two-step synthesis schemes result in the formation of SWCNT-dsDNA hybrid structures, as demonstrated by the increase in the average length, shown in the two distributions. However, the two-step reaction produces a significantly narrower length distribution, which likely reflects a higher yield of the desired SWCNT-dsDNA hybrid structure. This is reasonable, since in the one-step scheme, a percentage of the SWCNT segments

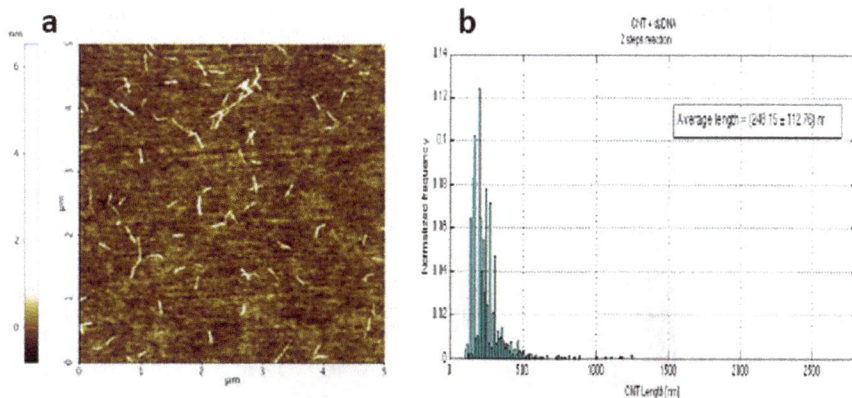

Figure 4 Tapping mode AFM image of the SWCNT-dsDNA solution generated by the two-step reaction. b) Histogram of the length distribution of the SWCNT-dsDNA hybrid structures obtained with the two-step reaction. The histogram was built by measuring the SWCNT length with ImageJ software. 1833 nanotube segments were measured, from different AFM images of different substrates on which the same solution was deposited. The average length and standard deviation of the CNT segments in the starting solution were found to be 248 $\pm$ 113 nm.

could be expected to be saturated with dsDNA on both ends, rendering them unable to link any further. The two-step scheme remedies this by introducing a new, unreacted population of activated SWCNTs which can bind with the saturated nanotubes.

Figure 5 shows a small area AFM scan of a dilute solution of SWCNT-dsDNA hybrids deposited on mica. Several different conformations can be observed, including both straight and kinked structures,which correspond to what would be expected for the SWCNT-dsDNA hybrids. Presumably, the kink could indicate the location of the dsDNA within the structure. We believe that these structures are well-suited for electrical characterization, the results of which will be reported separately.

3 Nanoassembly using DNA-PNA Hybrids

Peptide nucleic acid (PNA) is an analogue of DNA with a backbone made from N-(2-aminoethyl glycine) units instead of DNA's deoxyribose phosphate; the N-terminal of the PNA backbone corresponds to the 5' end of the DNA backbone [11, 12]. The last decade has seen increased attention paid and progress made in developing PNA targeting reagents and methods. PNA has been explored for uses in drug development and diagnostics [13]. Moreover, PNA binding to DNA can affect gene expression, and offers increased chemical stability in the presence of other enzymes. For example, PNA bound to a restriction enzyme site inhibits site cleavage by that enzyme [14]. PNA has also been explored for use with DNA origami as part of DX (double crossover)

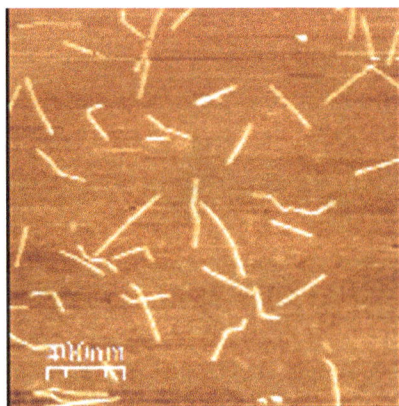

Figure 5   AFM scan of SWCNT-dsDNA Hybrids on Mica Prior to Electrode Deposition

molecules [15]. In this way, it has been incorporated into arrays of 2D nucleic acid tiles, a useful tool for self-assembling large-scale arrays on surfaces.

PNA can bind to DNA in several ways: duplex formation with ssDNA or RNA, triplex formation by Watson-Crick and Hoogsteen binding, and strand displacement for Watson-Crick binding of PNA base pairs to the DNA duplex [12]. PNA also forms double helical structures nearly identical to DNA. For this reason, design of complementary PNA labels is highly sequence and site-specific. In this work, we employ strand displacement of part of one strand of the DNA duplex in order to bind a PNA-DNA chimera to dsDNA [16]. Complementary base-pair binding can be used for DNA labeling; an identical backbone is not necessary for hybridization of labels by base pairs. Whereas DNA has a negatively charged backbone, the backbone of PNA is neutral. PNA-DNA duplexes are more stable at higher temperatures than DNA-DNA duplexes; the melting temperature of a PNA-DNA duplex increases in melting temperature by 1°C per base pair over a DNA-DNA duplex in the presence of 100mM NaCl [17]. Strand displacement/invasion of DNA by PNA takes advantage of the breathing modes inherent in heated DNA. When DNA is heated, the individual strands come apart slightly, allowing PNA to preferentially bind to the complementary base pairs. If PNA is introduced into the solution at this time, as the solution cools, the base pairs of the PNA strand will preferentially pair with complementary base pairs on the DNA, as there is less electrostatic repulsion as in DNA-DNA binding. The PNA forms a triplex, creating a loop in the DNA.

Stadler et al. proposed that naturally-occurring lambda DNA could be used as a scaffold for assembly of functional nanomaterials[16]. Toward that end, they demonstratedsolution-based binding of Au nanoparticle to synthetic double-stranded DNA via duplex invasion of a PNA-DNA chimera[16]. We sought to build upon this method by optimizing the binding conditions to control the placement of nanoparticles on a surface, with an eye toward ultimately using naturally-occurring λ-DNA in order to extend the length scales across which DNA-modulated architectures could be self-assembled. It is possible to tune the PNA sequence and incubation time and temperature in order to invade specific sites on -DNA, which measures 16 m in length[18, 19].This could enable site-selective placement of nanoscale architectures over length scales far exceeding the dimensions of DNA origami. One caveat is that the properties of differently-synthesized PNA greatly affect binding specificity and yield. Put simply, every different PNA designed may require a different binding process; there is no one-size-fits-all method of determining binding conditions for every PNA molecule.

To test the binding of PNA to lambda-DNA fragments, binding conditions were adapted from Chan et al and Zohar et al.[18, 19]. The PNA was a bis- with sequence: N-dig-OO-Lys-Lys-TCC TTC TC-OOO-JTJ TTJ JT-Lys-OO-Lys-O–COOH [18, 19]. The "J" base pair is pseudoisocytosine, which can replace cytosine in the sequence and reduces the binding sensitivity to ionic strength. This sequence preferentially invades DNA using triplex invasion [20]. Triplex invasion complexes are very stable at high temperatures. Lambda DNA (NEB, 48kbp) was digested with NcoI and XbaI for 1.5hrs at 37°C in NEB4 buffer, yielding a 607bp fragment. PNA (Biosynthesis, Inc., Lewisville, TX) was incubated with the restricted lambda for 25 hours at 37°C in the presence of 2mM NaCl, thenallowed to cool to room temperature on the benchtop. The gel was stained with SYBR green, an ultrasensitive dye, for imaging via 4% polyacrylamide gel-shift assay.

We also examined binding of PNA to a 45bp synthesized lambda DNA fragment [21]. The DNA sequence used was: 5' -GCA ACA GTG GCATGC ACC GAG AAG GAC GTT TGT AAT GTC CGCTCC-3', which is a fragment of lambda DNA from base pairs 24342–24386, with melting temperature 70.5°C.

PNA was mixed with single-stranded complementary and non-complementary strands in the following molar ratios to the single strands: 10:1, 5:1, 1:1, 0.5:1, and 0.1:1. Although we followed the protocols for binding PNA to DNA from Chan et al. [18] and Zohar et al. [19], The oligos were incubated with PNA for a shorter time because of their reduced size as compared to full lambda DNA. Briefly, PNA was added to DNA with 100 mM NaCl in 1X Tris-EDTA bufer and incubated in a water bath at 63°C for 1.5hrs. then allowed cool to room temperature. A 4 polyacrylamide gel was run at 90V for 2 hrs. We confirmed the binding of PNA to 45 base-pair dsDNA duplexes. In panel (a) of Figure 6, gel shifts can clearly be seen with PNA hybridized to DNA at molar ratios of 10:1 and 5:1. Gel-shift experiments with larger lambda fragments from restriction digests were less consistent. Experiments involving PNA bound to a restriction-digested enzyme were inconsistent, but it is thought that a gel-shift occurred with a 607bp fragment, as in Figure 6, panel (b).

Hybridization of PNA to DNA was performed following the methods of Stadler et al. [16]. In this method, 200 base-pair fragments of synthesized dsDNA are incubated with 25 base-pair long single-stranded (ss) PNA-DNA chimeras at 50 °C overnight, allowing the single-stranded PNA-DNA chimera to preferentially invade the charged DNA helix. Although this labeling was been done in solution, we sought to perform directed self-assembly of DNA structures on lithographically-patterned surfaces, increasing

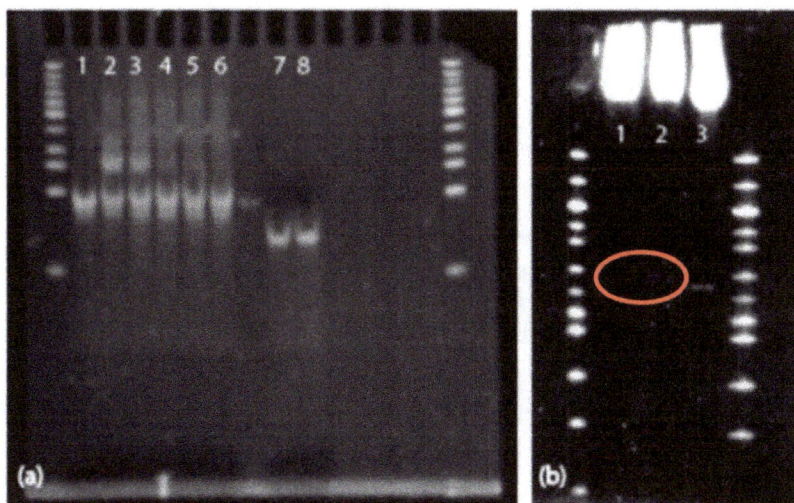

Figure 6 Gel-Shift Assays. PNA binds well to the 45bp dsDNA fragments at higher molar-ratios, but is barely visible (and not supershifted) on the larger restriction-enzyme-digested portion. (a) PNA gel-shift assay on 45bp dsDNA duplexes. Column 1 is 45bp dsDNA; 2 shows a gel shift due to PNA bound to DNA at a 10:1 molar ratio; 3 shows a gel shift due to PNA binding DNA at a 5:1 molar ratio. Columns 4, 5, and 6 show no gel shift; PNA was incubated with DNA at molar ratios of 1:1, 0.5:1, and 0.1:1, respectively. Column 7 shows 45bp ssDNA, and Column 8 is ssDNA incubated with 0.1:1 PNA:DNA (shows nobinding). (b) PNA gel-shift assay on 607bp fragment. Columns 1 and 2 shows gel shiftson 607bp lambda + PNA + an anti-digoxigenin Fab fragment and 607bp lambda + PNA, respectively. Column 3 shows the 607bp fragment alone.

nanopattern resolution and allowing self-assembly of functional nanomaterials on these surfaces. Placement of DNA molecules on a surface can be directed by binding them between patterned nanodots, which are functionalized with complementary oligonucleotides. This assembly scheme is illustrated in Figure 7.

The nanodots anchors were formed by nanoimprint lithography (NIL) and selective pattern transfer [22] on oxidized Si substrates. The Si NIL template was patterned by electron beam lithography using hydrogen silsesquioxane (HSQ) as a resist [23]. After annealing in air for densification, the HSQ itself served as the three-dimensional relief structure for NIL. Thermal NIL of poly-methylmethacrylate (PMMA) and subsequent pattern transfer processes yielded approximately spherical nanodots, with uniform diameter $\sim$7 – 8 nm. The nanodots were patterned in arrays of dimers with an inter-dot spacing of 60 nm.

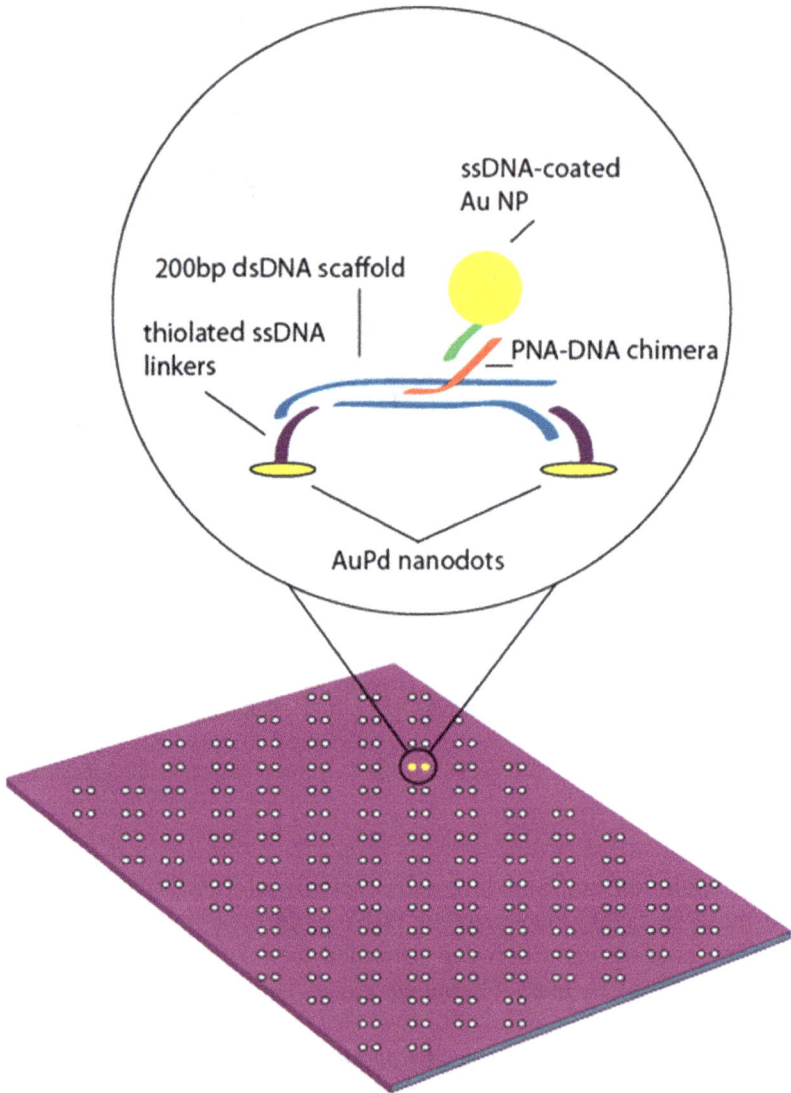

Figure 7   Schematic of Au NPs bound to PNA-DNA chimera - labeled dsDNA scaffold on AuPd nanodot-patterned surface.

The nanodots were functionalized with ssDNA using a thiol linker [24]. The following sequence was used: 5-ThioMC6-D / TTT TTTTTTTTTTT TAA CCT AAC CTT CAT; the 5' thiolated end bound to the AuPd dots. Prior to hybridization to the ssDNA linkers, duplex invasion of a PNA-DNA chimera was used to functionalize a 200bp fragment of dsDNA [c16]. Briefly, PNA was incubated with DNA in a 3X molar excess at 50 °C overnight in 10 mM phosphate buffer with 100mM NaCl. After overnight incubation, the resulting solution was left to cool to room temperature. Scaffolds were incubated with an equal molar concentration of linker arm ssDNA for several hours at room temperature.

The scaffold consisted of a 200bp dsDNA duplex with different length linker arms. The sequences of the two halves of the zigzag structure (shown in Fig. 8) were: 5-ACG TAC CAA ATA CGT CGA TTG GCT ACG TAA TAA CAA TTT CTA TTG GTT CCG CAA GCT GGC CCT CAC TTC AAC GCA **TTG TTA TTA** ATC TTC CAA TGG GCC ACC TAC CGT AGA CAC GGA CTC TCT ACG CGT TAT GCC TCA GCA TAT TGT TAT TAC TGC GGG ACA TAC GAT AGA GCT TTG CTA AAA TAA GTC CCT GCC TT-3 and 5-ACG TAC CAA ATA CGT CGA TTG GCT ACG TAA TAA CAA TTT CTA TTG GTG GAA AGG CAG GGA CTT ATT TTA GCA AAG CTC TAT CGT ATG TCC CGC AGT AAT AAC AAT ATG CTG AGG CAT AAC GCG TAG AGA GTC CGT GTC TAC GGT AGG TGG CCC ATT GGA AGA TTA ATA ACA ATG CGT TGA AGT GAG GGC CAG CTT GCG GA-3 with the PNA-binding sequence in bold. The sequence of the linker arms was: 5-ACG TAT TTG GTA CGT (T)n ATG AAG GTT AGG TTA-3, where poly-T chains of $n = 18$, 36, and 54 were inserted into the linker to extend their length and allow for more room for binding to occur.

The sequence of the PNA-DNA chimera, which bound to the bold section of the DNA, was: N-**TAA TAA CAA T**-Linker-CAC ATC TCT TCT GAA -3, with the PNA sequence in bold. After overnight incubation of ssDNA on nanodots, substrates were removed from hybridization buffer, dip-rinsed in a beaker of PBS, and partially dried with a laboratory wipe until only a thin film of PBS buffer remained on the substrate. The substrates were then

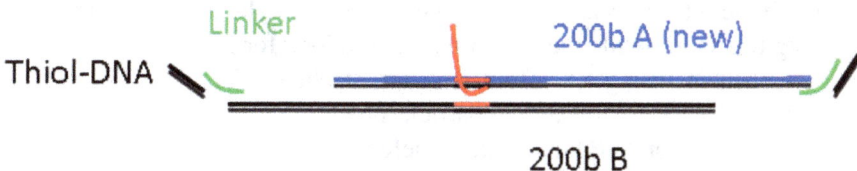

Figure 8 Schematic of the Zigzag DNA Scaffold, Including 3' Sticky Ends. The Linker Length was Varied to Optimize the Binding Percentage.

incubated with 20μL of 100–200n DNA-PNA scaffold solution and incubated motionless at room temperature for 6–8 hours. Following incubation of the PNA-containing 200bp scaffold, a 795nM solution of 10 nm gold nanoparticles coated with ssDNA complementary to the ssDNA portion of the PNA-DNA chimera was added. The sample was incubated overnight at room temperature to allow the nanoparticles to bind to the scaffold via the PNA-DNA chimera.

The binding yield was determined by SEM analysis. Approximately 500 or more sites for each sample were measured. Nanoparticles within a 30nm radius of the AuPd nanodots were deemed to be tethered by one end of the dsDNA duplex, while nanoparticles between the dots counted as trimers tethered by both ends of the dsDNA duplex. To determine the degree of non-specific nanoparticle binding to the surface, control experiments were performed with no PNA-DNA chimera attached to the scaffolds.

Linker arm lengths were varied in order to determine the optimal length for binding, and how much extra length was necessary to accommodate the nanoparticle binding to the PNA-DNA chimera and subsequent surface deposition. Poly-thymine linkers with lengths of 18, 36, and 54 base pairs were used. These made total scaffold lengths of 236, 272, and 290 base pairs, respectively. Table 5.1 illustrates results of experiments with varying linker lengths.

Stadler et al. were able to achieve a binding yield of ~26% in solution [16]. In analyzing surface binding to nanodots with 60nm spacing, we found that the best results came from a linker length involving the 36-T linker. In this case, 8.5% of dot-pairs had a gold nanoparticle in between them. An additional 13.4% of pairs had an AuNP nearby, indicating monovalent DNA-PNA scaffold tethering to a single AuPd nanodot. This contributes to a total of ~22% of dot pairs which have a nanoparticle self-assembled to the scaffold containing the DNA-PNA chimera, as shown in Fig. 9. This surface-based self-assembly result corresponds to nearly 85% of the solution yield. By contrast, only ~10% of dot pairs using scaffolds with 18-T and 54-T linkers displayed trimers or single-tethered nanoparticles. This may be explained by the 18-T linker being too short, while the 54-T linker is too long for efficient binding. Control experiments using a scaffold with no attached PNA-DNA chimera show ~6% of dot pairs with a nanoparticle nearby due to nonspecific Au nanoparticle binding or AuPd nanopattern defects.

Nanofabricated templates offer a starting point for directed self-assembly of double-stranded DNA arrays. As demonstrated here and elsewhere, site-specific labeling of dsDNA at tunable sites is possible using PNA, LNA, and nick-translation in solution [16, 18,19]. In this work, annealing PNA to

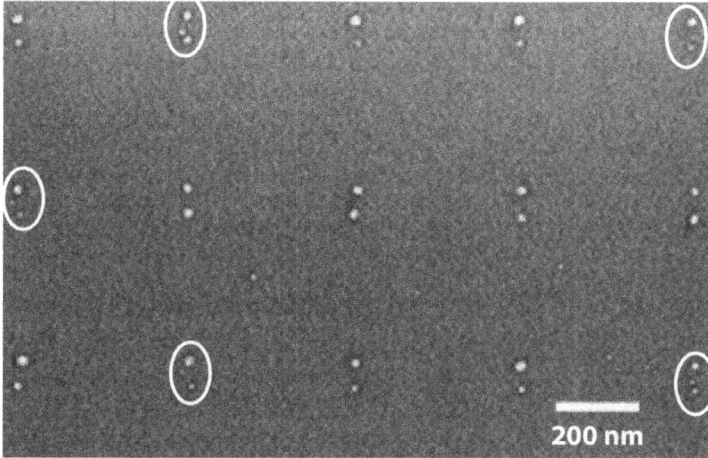

Figure 9 Directed assembly of PNA-labeled DNA scaffolds on AuPd dot pairs. Positions of trimers are highlighted by white ellipses.

DNA was done using only short (~45–600bp) double-stranded DNA fragments. Moreover, we have demonstrated surface-based self-assembly of gold nanoparticles onto a double-stranded DNA scaffold via PNA-DNA and DNA base-pair interactions. However, PNA invasion is extremely sensitive to conditions such as temperature, time, any linkers present (e.g. digoxigenin or other haptens) and base-pair sequence. Therefore, it remains difficult to massively tune hybridization conditions for PNA on lambda DNA. For DNA fragments up to 200 base pairs, PNA invasion works in solution for up to 26% of the sample [16]. Therefore, a rate of ~22% for surface-based assembly of nanoparticles in between and around AuPd nanodots constitutes 85% of solution-based binding efficiency. Labeling this DNA via PNA increases the resolution of these methods by enabling the site-specific placement of functional nano-objects (e.g. Au nanoparticles, quantum dots, nanowires, etc.) on surfaces. This has positive implications for self-assembly of larger, more complicated structures on surfaces.

4 Conclusions

In this work, we have demonstrated two approaches to integrating DNA with functional nanomaterials. One approach results in a hybrid nanostructure that has the potential to be further integrated into an active nanoelectronic device (i.e., where the DNA serves as the active element). The other approach demonstrates a new way to look at DNA as a scaffold for directed assembly. Together, both demonstrate the versatility of DNA as a technological material that can be exploited for many future applications.

Acknowledgements

We thank Oleg Gang, Andrea Stadler and Peter Sun at Brookhaven National Lab for their help, guidance and material support with PNA/DNA hybridization. We gratefully acknowledge financial support from the US-Israel Binational Science Foundation under Award SIBSF2006422 and the Office of Naval Research under Award N00014-09-1-1117. Additionalsupport from the Nanoscale Science and Engineering Initiative of the National Science Foundation under NSF Award CHE-0641523 and from the New York State Office of Science, Foundationunder NSF Award CHE-0641523 and from the New York State Office of Science, Technology, and Academic Research (NYSTAR) is also gratefully acknowledged.

References

[1] F. A. Aldaye, A. L. Palmer, and H. F. Sleiman, Assembling Materials with DNA as the Guide, *Science*, **321**(5897): 1795–1799 (2008).
[2] J. Fu and H. Yan, Controlled drug release by a nanorobot, *Nat Biotech,* **30**(5): 407–408 (2012).
[3] L. M. Adleman, Computing with DNA, *Sci Am*, **279**(2): 54–61 (1998).
[4] J. C. Genereux and J.K. Barton, Mechanisms for DNA Charge Transport, *Chem Rev*, **110**(3):1642–1662 (2010).
[5] X. Guo, J. P. Small, J. E. Klare, Y. Wang, M. S. Purewal, I. W. Tam, B. H. Hong, R. Caldwell, L. Huang, S. O'Brien, J. Yan, R. Breslow, S. J. Wind, J. Hone, P. Kim, and C. Nuckolls, Covalently bridging-gaps in single-walled carbon nanotubes with conducting molecules, *Science*, **311**(5759): 356–359 (2006.).
[6] X. F. Guo, A. Whalley, J. E. Klare, L. M. Huang, S. O'Brien, M. Steigerwald, and C. Nuckolls, Single-molecule devices as scaffolding for multicomponent nanostructure assembly, *Nano Lett,*. **7**(5):1119–1122 (2007).
[7] X. F. Guo, A. A. Gorodetsky, J. Hone, J. K. Barton, and C. Nuckolls, Conductivity of a single DNA duplex bridging a carbon nanotube gap. *Nature Nanotechnology*, **3**(3): 163–167 (2008).

[8] X. Y. Huang, R. S. McLean, and M. Zheng, High-resolution length sorting and purification of DNA-wrapped carbon nanotubes by size-exclusion chromatography, *Anal Chem*, **77**(19): 6225–6228 (2005).

[9] DNA strands were purchased from Syntezza.

[10] Y. Weizmann, D. M. Chenoweth, and T. M. Swager, Addressable Terminally Linked DNA-CNT Nanowires, *J Am Chem Soc*, **132**(40):14009–14011 (2010).

[11] T. Bentin and P. E. Nielsen, In vitro transcription of a torsionally constrained template, *Nucleic Acids Res*, **30**(3): 803–809 (2002).

[12] P. E. Nielsen, M. Egholm, and O. Buchardt, Peptide Nucleic-Acid (PNA) - a DNA Mimic with a Peptide Backbone, *Bioconjugate Chem*, 1994. **5**(1): p. 3–7.

[13] P. E. Nielsen, A new molecule of life? Peptide nucleic acid, a synthetic hybrid of protein and DNA, could form the basis of a new class of drugs- and of artificial life unlike anything found in nature, *Sci Am*, **299**(6): 64–71 (2008).

[14] P. E. Nielsen, M. Egholm, R. H. Berg, and O. Buchardt, Sequence Specific-Inhibition of DNA Restriction Enzyme Cleavage by PNA, *Nucleic Acids Res*, 1993. **21**(2): 197–200.

[15] P. S. Lukeman, A. C. Mittal, and N. C. Seeman, Two dimensional PNA/DNA arrays: estimating the helicity of unusual nucleic acid polymers, *Chem Commun*, (15): 1694–1695 (2004).

[16] A. L. Stadler, D.Z. Sun, M. M. Maye, D. van der Lelie, and O. Gang, Site-Selective Binding of Nanoparticles to Double-Stranded DNA via Peptide Nucleic Acid "Invasion", *ACS Nano*, **5**(4): 2467–2474 (2011).

[17] Applied biosystems - support. https://www2.appliedbiosystems.com /support/seqguide.cfm?

[18] E. Y. Chan, N. M. Goncalves, R. A. Haeusler, A. J. Hatch, J. W. Larson, A. M. Maletta, G. R. Yantz, E.D. Carstea, M. Fuchs, G.G. Wong, S.R. Gullans, and R. Gilmanshin, DNA mapping using microfluidic stretching and single-molecule detection of fluorescent site-specific tags, *Genome Res*, **14**(6): 1137–1146 (2004).

[19] H. Zohar, C. L. Hetherington, C. Bustamante, and S. J. Muller, Peptide Nucleic Acids as Tools for Single-Molecule Sequence Detection and Manipulation, *Nano Lett*, **10**(11): 4697–4701 (2010).

[20] P. E. Nielsen, Peptide nucleic acids : methods and protocols. Methods in molecular biology., Totowa, N. J.: Humana Press. 274 p (2002).

[21] Integrated DNA Technologies, Coralville, IA

[22] M. Schvartzman and S.J. Wind, *Robust Pattern Transfer of Nanoimprinted Features for Sub-5-nm Fabrication.* Nano Lett, **9**(10): 3629–3634 (2009).

[23] H. Namatsu, Y. Takahashi, K. Yamazaki, T. Yamaguchi, M. Nagase, and K. Kurihara, Three-dimensional siloxane resist for the formation of nanopatterns with minimum linewidth fluctuations, *J Vac Sci Technol B*, **16**(1): 69–76 (1998).

[24] M. Palma, J. J. Abramson, A. A. Gorodetsky, E. Penzo, R. L. Gonzalez, M. P. Sheetz, C. Nuckolls, J. Hone, and S. J. Wind, Selective Biomolecular Nanoarrays for Parallel Single-Molecule Investigations, *J Am Chem Soc*, **133**(20): 7656–7659 (2011).

Biographies

Shalom J. Wind received his Ph.D. in Physics from Yale University in 1987. Following his doctoral studies, he worked at IBM's Thomas J. Watson Research Center, focusing primarily on nanoelectronic devices. He moved to the Department of Applied Physics and Applied Mathematics at Columbia University in 2003. Dr. Wind's present research focuses on molecular-scale fabrication and the interface between biological and technological materials and systems.

Erika Penzo was born in Schio, Italy, in 1983. She got her BSc and MSc in Physics Engineering from Politecnico di Milano, Italy. She is currently a PhD candidate at Columbia University (New York) in the Applied Physics department. Her research focuses on nanofabrication, self-assembly and DNA nanotechnology.

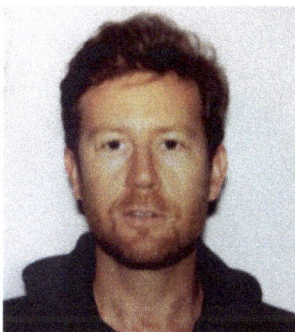

Matteo Palma received his Ph.D. in Physical Chemistry from University Louis Pasteur in 2007. He was a post-doctoral fellow at Columbia University from 2008–2012 in the Departments of Mechanical Engineering and Applied Physics and Applied Mathematics. Dr. Palma will take up a position as a lecturer at Queen Mary College, University of London, beginning in September 2013. Dr. Palma's research interests include the controlled self-assembly of functional nanostructures on surfaces with true nanoscale resolution, with a particular focus on supramolecular interactions to drive the self-organization of nano-moieties on (nanopatterned) substrates.

The Many Faces of Diphenylalanine

Mohtadin Hashemi, Peter Fojan and Leonid Gurevich*

*Institute of Physics and Nanotechnology, Aalborg University,
9220 Aalborg East, Denmark*
*Corresponding author: lg@nano.aau.dk

Received 18 June 2013; Accepted 06 August 2013; Publication 6 August 2013

Abstract

Diphenylalanine is well known to form complex self-assembled structures, including peptide nanowires, with morphologies depending on N- and C-terminal modifications. Here we report that significant morphological variations of self-assembled structures are attainable through pH variation of unmodified diphenylalanine in trifluoroethanol. The obtained self-assembled diphenylalanine nanostructures are found to vary drastically with pH, incubation time, and diphenylalanine concentration in solution. The observed structures ranged from structured films at neutral and alkaline conditions to vertically aligned nanowires and sponge-like structures at acidic conditions. These observations are corroborated by the results of electrostatic modelling, indicating the disappearance of the dipole moment at high pH values. This also emphasizes the importance of the dipole moment for the resulting self-assembled structures. Our results suggest that, in comparison to the commonly described procedure of diphenylaniline nanowire growth through aniline vapor treatment, strictly anhydrous conditions are not necessarily required.

Keywords: Diphenylalanine, peptide self-assembly, peptide nanowires, peptide nanotubes.

1 Introduction

Self-assembly is a promising route to form functional nanostructured materials. In particular, peptides are specifically designed by Nature to self-assemble

Journal of Self-Assembly and Molecular Electronics, Vol. 1, 195–208.
doi: 10.13052/jsame2245-4551.123

into complex structures [1]. This makes peptides a viable building material, in particular when biocompatibility is required [2–4]. Another major technological advantage of peptides is the well-known coupling chemistry and the diversity of chemical and physical properties of individual amino acids leading to a vast number of possible combinations [5].

From a different perspective, although biological functions have been generally ascribed to proteins, a vast number of biologically active short peptides have been characterized in the past with functions ranging from storage and transport to antimicrobial activity [6–9] and toxins [10, 11]. Recently, short peptides have been linked to a number of neurodegenerative disorders such as Alzheimer's, Parkinson's, and Prion diseases [12–14]. Studies have revealed that these conditions are related to peptide self-assembly properties, commonly known as amyloidogenesis. During this process a soluble and innocuous peptide turns into insoluble aggregates forming amyloid fibrils [14–16].

The search for a model system to study the mechanism of amyloid formation led to the shortest possible self-assembling peptide consisting of just two units - diphenylalanine (FF) (Figure 1) [17, 18]. Further studies have demonstrated that FF can assemble into a variety of structures including peptide nanotubes (PNTs) [19].

The role of electrostatic interaction between FF units during PNT growth has been investigated using FF with differently modified termini, which have been found to form vertically aligned tubes, non-aligned tubes, or only a minimal amount of tubes [19].

Another route for the formation of FF peptide nanowires (PNW) is based on aniline vapor treatment of FF amorphous films at elevated temperatures [20, 21]. Furthermore, PNWs can serve as templates for the formation of more complex structures for various applications, e.g. PNW/polyaniline core/shell

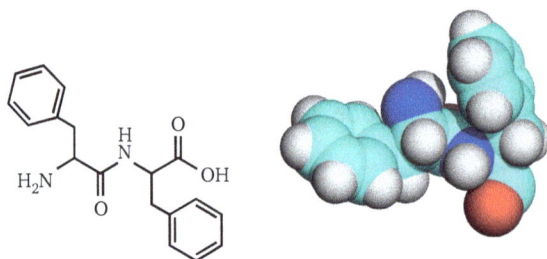

Figure 1   2D and 3D (CPK model) structure of diphenylalanine.

[22], coaxial nanocables [23], stencil masks for nanolithography [24], and electrochemical sensors [25–27] have been demonstrated.

In the present paper we explore the possibility to affect the morphology of self-assembled structures by tuning the charge on the peptide via pH adjustment. We report the formation of various self-assembled structures using solutions of FF in TFE at different pH conditions and different incubation times.

2 Experimental Procedures

2.1 Synthesis of Diphenylalanine

Diphenylalanine was synthesized ($0.1 mmol$ theoretical yield) using SPPS on an Activo P11 Automated Peptide Synthesizer (Activotec, UK) following the protocol described in Table 1.

DIPEA (N,N'-diisopropylethylamine), HBTU (O-(Benzotriazol-1-yl)-N,N,N',N'-tetramethyluronium hexafluorophosphate), piperidine, and Fmoc-Phe-OH were acquired from Advanced Chemtech. Phe on Wang resin ($0.75 mmol/g$), HOBt (N-hydroxybenzotriazole), DCM (dichloromethane), and DMF (N,N'-dimethylformamide) were obtained from Iris Biotech GmbH.

Table 1 Solid phase peptide synthesis protocol

Procedure	Reagent	Time [min]	Cycles [#]
Loading	Fmoc protected Phe on Wang resin		
Swelling	7 mL DMF	15	
	7 mL DMF	45	
Deprotection	7 mL 25 % Piperidine	3	
	7 mL 25 % Piperidine	13	
Washing	7 mL DMF	1	5
Amino acid	Fmoc-Phe-OH (1.75 mmol) in 3.5 mL 0.5 M HBTU/HOBt	1	
Activation of amino acid	3.5 mL, 1 M DIPEA	1	
Coupling		45	
Washing	7 mL DMF	1	4
Deprotection	7 mL 25 % Piperidine	3	
	7 mL 25 % Piperidine	13	
Washing	7 mL DMF	1	5
Washing	7 mL DCM	1	3

Cleavage

Removal of side-chain protection groups and cleavage of FF from Wang resin was achieved by 90 min incubation in $4mL$ solution, consisting of 95% TFA (trifluoroacetic acid) (Iris Biotech GmbH), 2.5% deionized water, and 2.5% TIS (triisopropylsilane) (Fluka). After cleavage, FF in TFA/TIS was upconcentrated in a rotary evaporator and subsequently precipitated using ice-cold diethylether (Iris Biotech GmbH). The obtained FF powder was dissolved in 0.1% TFA and freeze dried.

2.2 Self-assembly of Diphenylalanine

Lyophilized FF was dissolved in TFE (Fluka), and ultrasonicated for 1 min following the initial disolution step. For pH adjustment, either $50\mu L$ $1M$ NaOH (Bie & Berntsen) or $50\mu L$ $1M$ HCl (Fluka) was added to $100\mu L$ of the FF solution. The FF solution was equilibrated at ambient temperature for 5, 10, or 30 minutes. To avoid pre-aggregation, fresh stock solutions were prepared for each set of experiments.

$10x10mm$ Si-wafers coated with $300nm$ of thermally grown SiO_2 (NOVA wafers) were cleaned by ultrasonication in 99% ethanol (Kemetyl), rinsed and dried under a stream of N_2.

For self-assembly, a $20\mu L$ droplet of FF solution was deposited on the Si-wafer and allowed to completely evaporate at ambient temperature.

To grow PNWs, self-assembled peptide films on Si-substrates were vapor treated with aniline (Sigma-Aldrich) in vacuum at 100^oC.

2.3 SEM

Samples were imaged using a Zeiss EVO 60 with an acceleration voltage of $10kV$ and samples used "as is", unless otherwise stated. If coating was necessary, the sample has been sputter coated with Au for 30 seconds using Edwards S150B sputter coater.

2.4 Electrostatic Map

The diphenylalanine structure obtained by energy minimization (steepest descent and Newton-Raphson) was fed to the Kryptonite PDB2PQR server [28], with application of PROPKA [29] for pKa calculations. Adaptive Poisson-Boltzmann Solver (APBS) [30] was used to calculate the electrostatic potential using the linearized Poisson-Boltzmann equation. Parameters, except pH, for Kryptonite server and APBS were kept at default values. To account for the presence of residual water in TFE, the dielectric constant was set to 10.

3 Results & Discussion

Elecrostatic modelling of the diphenylalanine peptide revealed three distinct states: at pH 0–2 the peptide is cationic, between pH 3–8 it is zwitterionic, and above pH 8 it is anionic (Figure 2).

Diphenylalanine was found to initially form an amorphous film on the substrate, regardless of concentration or incubation time (Figure 3). The observed film thickness scaled linearly. Contrary to the claims of [22], stating that anhydrous environment is required to form film, we found that peptide film formed under ambient conditions and in Ar atmosphere as well.

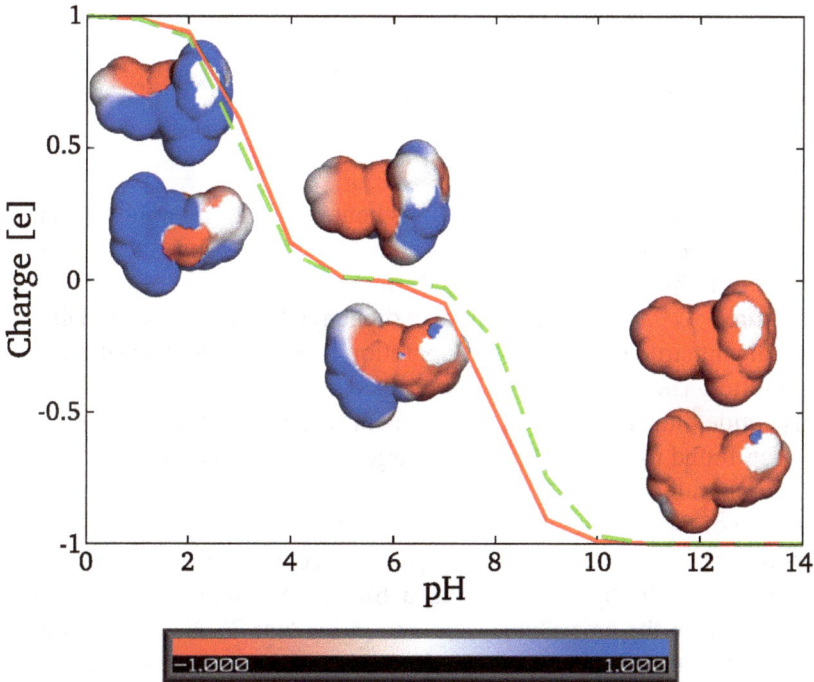

Figure 2 Titration curve and electrostatic potential maps projected onto the solvent accessable surface of FF peptide at pH 0, 7, and 14. The colorbar is in units $\frac{k_B T}{e}$ at $298K$. Solid line represents the charge of the folded state of FF, while dashed line corresponds to the unfolded state (states derived by PROPKA).

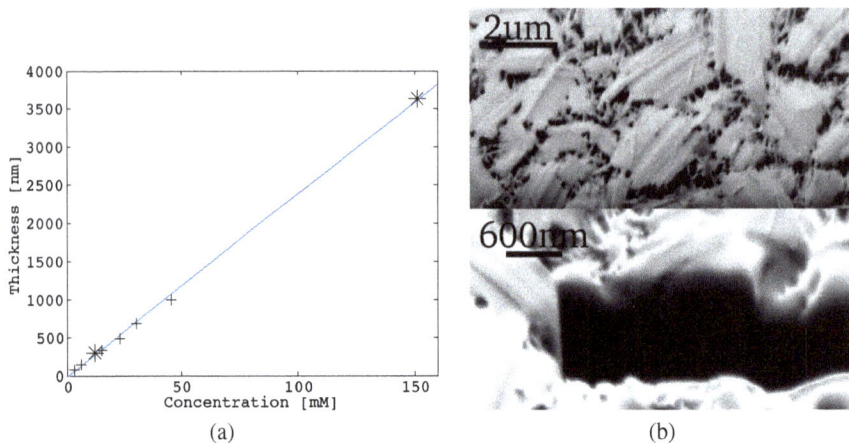

(a)   (b)

Figure 3   (a) Thickness of the peptide film formed on the substrate under ambient conditions plotted versus FF concentration. Data from the present study denoted as *, data from [21] marked as +. The obtained relation is linear with the best fit $d = 24[FF] - 29$, where the FF concentration is expressed in mM and the thickness d is obtained in nm. (b) SEM image of the peptide film formed in ambient environment, using $15mM$ FF in TFE and $5min$ incubation. Bottom shows peptide film milled with FIB *in situ*. Images taken with Zeiss 1540 XB with acceleration voltage of $1kV$.

According to literature the peptide film is believed to be amorphous [31], however in our case the film was not a continuous entity, but rather consisted of many individual, interconnected, micro-crystallites of typical size 2–$3\mu m$ (Figure 3). The observed behavior of diphenylalanine is most probably related to solvent structuring during evaporation leading to the formation of microcrystallites. The network of interconnected micro-crystallites forms the observed peptide film. This is in good agreement with [32], where the peptide film has been found to be an assembly of large bundles of tubular structures arranged as spherulites.

Incubation of the peptide film in aniline vapor at $100^{\circ}C$ for $8hrs$ yielded PNWs, with size and density depending on FF concentration (Figure 4). The peptide film, initially formed from a $6mM$ FF solution, underwent rearrangement from the amorphous state into individual PNWs (Figure 4.a) upon treatment with aniline vapor. No distinctive growth direction or size distribution was observed; the largest PNWs observed were in the range of 360–$420nm$ in diameter. Doubling the FF concentration to form the initial peptide film resulted in a similar behavior as described above, except for a higher wire density.

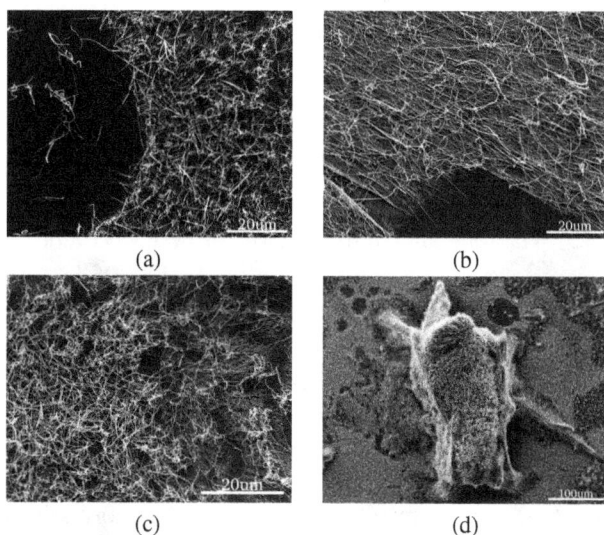

Figure 4 SEM images of different peptide films treated with aniline vapor at $100^{\circ}C$ for $8hrs$. The precursor film was obtained using (a) $6mM$, (b) $12mM$, and (c) $150mM$ solutions of FF in TFE. (d) Zoom out showing a 3D island on peptide film from (c).

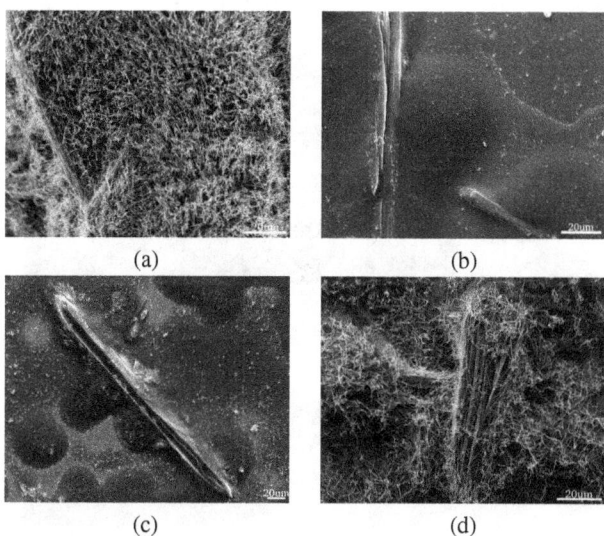

Figure 5 (a) Zoomed image of the island in Figure 4.d showing vertically aligned PNWs. Peptide films were formed from $150mM$ FF solution and subsequently incubated in aniline vapor at $100^{\circ}C$ for (b) 2, (c) 4, and (d) 8 hours.

The peptide film formed from a solution with $150mM$ FF exhibited much higher density of PNWs as compared to the $6mM$ and $12mM$ samples. Besides that, the $150mM$ samples contained several islands (Figure 4.d and Figure 5) of vertically aligned wires. Preliminary results suggest that while formation of such 3D islands is independent of the incubation time in aniline vapor - the images obtained after 2, 4, and 8 hours incubation, all show similar structures (Figure 5). The total area of the islands formed was found to be time dependent.

Acidic conditions, independent of incubation time, tend to favor the formation of 3D structures (Figure 6). The height profiles of the 3D structures varied between $90–130\mu m$ for smaller structures and up to $210–270\mu m$ for the largest structures observed. Incubation for 5 and 10 minutes yielded fibrous 3D "mushroom" structures (Figure 6.c). Upon extended incubation ($30min$), we observed merging of the fibrous 3D "mushroom" structures into porous "sponges" (Figure 6.d). These results suggest that the morphology of the 3D structures is incubation time dependent. Fibrous mushroom structures were found to form after incubation times of 5 and 10 minutes, while porous sponge-like structures require 30 minutes of incubation.

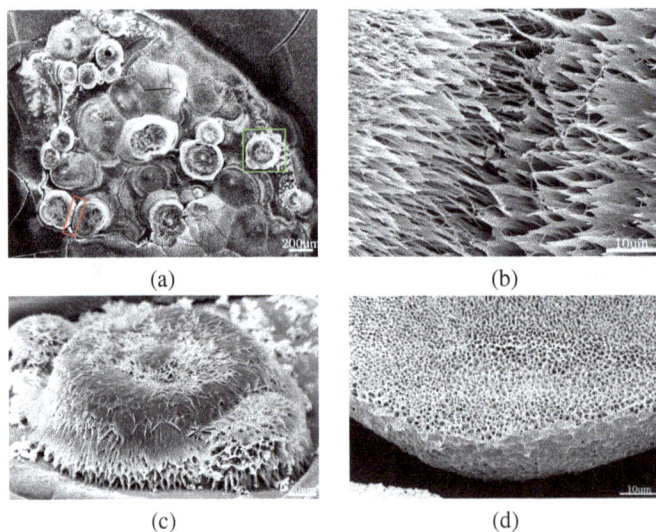

(a)   (b)   (c)   (d)

Figure 6   (a) SEM image showing 3D islands on an FF film formed from a $150mM$ FF solution under acidic conditions and $5min$ incubation. The highlighted regions are magnified in panels (b) and (c). (d) Close-up of a sponge-like 3D structure formed after $30min$ incubation.

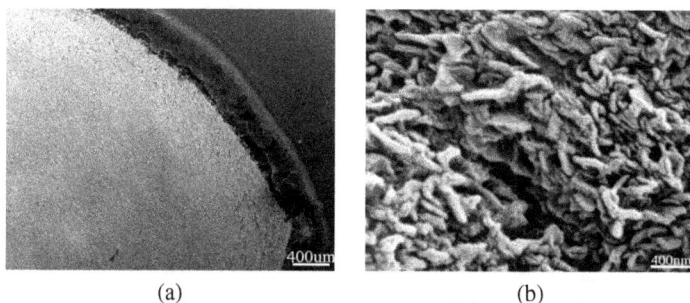

(a)   (b)

Figure 7   SEM images of self-assembled structures obtained under alkaline conditions from a $150mM$ FF solution deposited at ambient conditions. Panel (b) shows a zoom-in on the surface structure. Samples were coated with Au prior to SEM imaging.

Structure formed under alkaline conditions had a drastically altered morphology (Figure 7). No PNW growth was observed and the structure consisted of small rod- and flake-like structures interwoven in a complex manner independent of the incubation time. This observation is in line with the electrostatic modelling results showing a lack of a distinct dipole moment at pH values above 8 indicating the importance of a dipole moment for structured self-assembly of diphenylalanine.

4 Conclusion

All the reports on diphenylalanine self-assembly known in literature employed hexafluoro-2-propanol (HFIP) as a solvent. Moreover, strictly anhydrous conditions have been found to be necessary to achieve peptide nanowire growth upon aniline vapor treatment [20]. Our results suggest that a less hydrophobic solvent, trifluoroethanol, is capable of inducing the self-assembly of diphenylalanine. Moreover, by varying the pH of the trifluoroethanol solution of unmodified diphenylalanine, significant morphological variations were observed. While neutral and alkaline conditions favor formation of microstructured films, vertically aligned nanowires and sponge-like structures were obtained at acidic conditions. This observation is in line with the results of [19], where a peptide with positively charged termini has been found to form well aligned peptide nanotubes, while negatively charged peptide have not formed any aligned structures. This has been attributed to the peptide interaction with a negatively charged silicon surface used as a substrate. In contrary to [19] we did not observe assembly into peptide nanowires at neutral pH, however this can be related to a slightly different protonation state of our substrate and peptide molecules as compared to [19]. Moreover, we found that using trifluoroethanol as a solvent did not require strictly anhydrous conditions to initiate the PNW growth through aniline vapor treatment. Similar results were observed in argon and ambient atmosphere. At lower concentrations the PNWs were observed to grow along the substrate, while at higher concentrations, islands of vertically aligned nanotubes started to appear. These results demonstrate a much wider parameter space for growing diphenylalanine nanowires and shows that pH can be used as an effective tool to control the morphology of peptide self-assembled structures [33].

Acknowledgements

The authors gratefully acknowledge financial support from the Obel Family Foundation.

References

[1] A. M. Kushner and Z. Guan. Modular Design in Natural and Biomimetic Soft Materials. *Angew. Chem. Int. Ed.*, **50**:9026–9057 (2011).

[2] L. Liu, K. Busuttil, S. Zhang, Y. Yang, C. Wang, F. Besenbachera and M. Dong. The role of self-assembling polypeptides in building nanomaterials. *Phys. Chem. Chem. Phys.*, **13**:17435–17444 (2011).

[3] R. de la Rica and H. Matsui. Applications of peptide and protein-based materials in bionanotechnology. *Chem. Soc. Rev.*, **39**:3499–3509 (2010).

[4] N. Srinivasan and S. Kumar. Ordered and disordered proteins as nanomaterial building blocks. *WIREs Nanomed Nanobiotechnol*, **4**:204–218 (2012).

[5] R. B. Merrifield. Solid Phase Peptide Synthesis. I. The Synthesis of a Tetrapeptide. *J. Am. Chem. Soc.*, **85**(14):2149–2154 (1963).

[6] T. Ganz. Defensins: Antimicrobial peptides of innate immunity. *Nature Reviews Immunology*, **3**:710–720 (2003).

[7] E. Guani-Guerra , T. Santos-Mendoza, S. O. Lugo-Reyes and L. M. Teran. Antimicrobial peptides: General overview and clinical implications in human health and disease. *Clinical Immunology*, **135**:1–11 (2010).

[8] G. Maroti, A. Kereszt, E. Kondorosi and P. Mergaert. Natural roles of antimicrobial peptides in microbes, plants and animals. *Research in Microbiology*, **162**(4):363–374 (2011).

[9] M. Zasloff. Antimicrobial peptides of multicellular organisms. *Nature*, **415**:389–395 (2002).

[10] H. Raghuraman and A. Chattopadhyay. Melittin: a Membrane-active Peptide with Diverse Functions. *Biosci Rep*, **27**:189–223 (2007).

[11] G. Terlau and B. M. Olivera. *Conus* Venoms: A Rich Source of Novel Ion Channel-Targeted Peptides. *Physiol Rev*, **84**:41–68 (2004).

[12] G. Bhak and Y.-J. Choe and S. R. Paik. Mechanism of amyloidogenesis: nucleation-dependent fibrillation versus double-concerted fibrillation. *BMP reports*, **42**(9):541–551 (2009).

[13] J. A. Hardy and G. A. Higgins. Alzheimer's disease: The amyloid cascade hypothesis. *Science*, **256**:184–185 (1992).

[14] M. Stefani, C. M. Dobson. Protein aggregation and aggregate toxicity: new insights into protein folding, misfolding diseases and biological evolution. *Journal of Molecular Medicine*, **81**(11):678–699.

[15] J. W. Kelly. The alternative conformations of amyloidogenic proteins and their multi-step assembly pathways. *Curr. Opin. Struct. Biol.*, **8**(1):101–106 (1998).

[16] B. H. Toyama and J. S. Weissman. Amyloid Structure: Conformational Diversity and Consequences. *Annu. Rev. Biochem.*, **80**:557–85 (2011).

[17] A. T. Petkova, R. D. Leapman, Z. Guo, W.-M. Yau, M. P. Mattson, R. Tycko. Self-Propagating, Molecular-Level Polymorphism in Alzheimer's β-Amyloid Fibrils. *Science*, **307**:262–265 (2005).

[18] M. Reches and E. Gazit. Casting Metal Nanowires Within Discrete Self-Assembled Peptide Nanotubes. *Science* **300**:625–627 (2003).

[19] M. Reches and E. Gazit. Controlled patterning of aligned self-assembled peptide nanotubes. *Nature Nanotechnology*, **1**:195–200 (2006).

[20] J. Ryu and C. B. Park. High-Temperature Self-Assembly of Peptides into Vertically Well-Aligned Nanowires by Aniline Vapor. *Advanced Materials*, **20**:3754–3758 (2008).

[21] J. Ryu and C. B. Park. Solid-Phase Growth of Nanostructures from Amorphous Peptide Thin Film: Effect of Water Activity and Temperature. *Chemistry of Materials*, **20**:4284–4290 (2008).

[22] J. Ryu and C. B. Park. Synthesis of Diphenylalanine/Polyaniline Core/Shell Conducting Nanowires by Peptide Self-Assembly. *Angew Chem Int Edit*, **48**:4820–4823 (2009).

[23] O. Carny, D. E. Shalev and E. Gazit. Fabrication of Coaxial Metal Nanocables Using a Self-assembled Peptide Nanotube Scaffold. *Nano Letters*, **6**(8):1594–1597 (2006).

[24] M.B. Larsen, K.B. Andersen, W.E. Svendsen and J. Castillo-Leon. Self-Assembled Peptide Nanotubes as an Etching Material for the Rapid Fabrication of Silicon Wires. *BioNanoSci.*, **1**:31–37 (2011).

[25] I. De Oliveira Matos and W. A. Alves. Electrochemical Determination of Dopamine Based on Self-Assembled Peptide Nanostructure. *ACS Appl. Mater. Interfaces* **3**:4437–4443 (2011).

[26] M. Yemini, M. Reches, J. Risphon and E. Gazit. Novel Electrochemical Biosensing Platform Using Self-assembled Peptide Nanotubes. *Nano Letters*, **5**(1):183–186, 2005.

[27] J. Yuan, J. Chen, Z. Wu, K. Fang and L. Niu. A NADH biosensor based on diphenylalanine peptide/carbon nanotube nanocomposite. *Journal of Electroanalytical Chemistry*, **656**:120–124 (2011).

[28] T. J. Dolinsky, J. E. Nielsen, J. A. McCammon and N. A. Baker. PDB2PQR: an automated pipeline for the setup, execution, and analysis of Poisson-Boltzmann electrostatics calculations. *Nucleic Acids Research*, **32**:W665-W667 (2004).

[29] M. H. M. Olsson, C. R. Sndergard, M. Rostkowski and J. H. Jensen. PROPKA3: Consistent Treatment of Internal and Surface Residues in Empirical pKa predictions. *Journal of Chemical Theory and Computation*, **7**(2):525–537 (2011).

[30] N. A. Baker, D. Sept, S. Joseph, M. J. Holst and J. A. McCammon. Electrostatics of nanosystems: application to microtubules and the ribosome. *Proc. Natl. Acad. Sci. USA*, **98**:10037–10041 (2001).
[31] J. Ryu and C. B. Park. High Stability of Self-Assembled Peptide Nanowires Against Thermal, Chemical, and Proteolytic Attacks. *Biotechnol Bioeng*, **105**(2):221–230 (2010).
[32] N. Hendler, N. Sidelman, M. Reches, E. Gazit, Y. Rosenberg and S. Richter. Formation of Well-Organized Self-Assembled Films from Peptide Nanotubes. *Advanced Materials*, **19**:1485–1488 (2007).
[33] L. Gurevich, T. W. Poulsen, O. Z. Andersen, N. L. Kildeby and P. Fojan. PH-dependent self-assembly of the short Surfactant-like peptide KA_6. *J. Nanosci. Nanotechnol.*, **10**:7946–7950 (2010).

Biographies

Mohtadin Hashemi received his Master's degree in Nanobiotechnology from the Institute of Physics and Nanotechnology, Aalborg University in 2013. His research interests are focused on protein interactions, molecular modelling and functional surfaces.

Peter Fojan received his Ph.D. in Biotechnology at the University of technology Graz, Austria in 1997. He initially worked on industrial genetics of eukaryotic organisms. During his postdoc time at Aalborg University at the Department of Biotechnology he moved into the area of protein physics and molecular modelling. With the startup of Nanotechnology at AAU he moved to the Department of Physics and Nanotechnology where he became an Associate Professor in 2009. His research interests are centered around biological and small molecules and their interactions with cells and surfaces in general, for medical, sensor applications and as antibacterial agents.

Leonid Gurevich received his Ph.D. in Physics at the Institute of Solid State Physics (Chernogolovka, Russia) in 1994. He initially worked on high-Tc superconductors but during his postdoc stay at Delft University of Technology became excited about nanotechnology and possibility of charge transport through a single molecule. Since 2005 he is an Associate Professor at Aalborg University. His research interests are focused on molecular electronics, biosensors, functional surfaces and nanofabrication.

Experimental Investigation of Self-Assembled Opal Structures by Atomic Force Microscopy, Spectroscopic Ellipsometry and Reflectometry

Natalia Alekseeva[1], Grigory Cema[1], Aleksey Lukin[1], Svetlana Pan'kova[1], Sergei Romanov[2], Vladimir Solovyev[1], Victor Veisman[1] and Mikhail Yanikov[1]

[1]*Department of Physics, Faculty of Physics and Mathematics, Pskov State University, Lenin Square 2, 180000 Pskov, Russia;*
e-mail: kaf-phy@psksu.ru
[2]*Institute of Optics, Information and Photonics, University of Erlangen-Nuremberg, Haber Str. 9a, 91058 Erlangen, Germany; Ioffe Physical Technical Institute, Polytechnicheskaya Street, 26,194021 St. Petersburg, Russia,*
e-mail: Sergei.Romanov@mpl.mpg.de

Received 19 September 2013, Accepted 15 November 2013,
Published 10 December 2013

Abstract

Self-assembled opal crystals (bulk silica opals and PMMA thin opal films) have been studied by atomic force microscopy (AFM) and optical spectroscopy. Reflectance and transmittance spectra ($R(\lambda)$ and $T(\lambda)$, respectively) as well as spectra of ellipsometric parameters $\Psi(\lambda)$ and $\Delta(\lambda)$ demonstrate pronounced changes with changing the angle of light incidence. Diameters of spheres obtained from AFM-images correspond to those obtained from Bragg fit to the diffraction resonance dispersions. The band of light losses detected by ellipsometry at the spectral range of avoided band crossing of opal eigenmodes was assigned to the energy exchange between these modes.

Keywords: opal, photonic crystal, Bragg diffraction, photonic bandgap structure, reflectance and transmittance spectra, ellipsometry, atomic force microscopy.

Journal of Self-Assembly and Molecular Electronics, Vol. 1, 209–222.
doi: 10.13052/jsame2245-4551.124

1 Introduction

In the last decades opal crystals attracted considerable attention as nanoporous templates for the fabrication of various composite materials. Closely packed silica or poly-methyl methacrylate (PMMA) beads of self-assembled synthetic opal (see the surface image in Fig. 1) form the face centered cubic (FCC) lattice [4]. Since the diameters D of beads are in the range of hundreds nanometers, opals are widely used as 3-dimensional photonic crystals for the visible [1, 2, 3].

Optical properties of 3-dimensional opal crystals are usually characterized by the angle-resolved reflectance and transmittance spectroscopy, whereas their structure is examined by scanning electronic microscopy (SEM) and atomic force microscopy (AFM). Considering the high costs of SEM equipment, the SPM methods in combination with digital processing may become prospective and competitive for express investigation of opal-based materials [5].

Among other optical techniques, ellipsometry is designed to provide such characteristics as the refractive index n and the extinction coefficient k of the material under study. Ellipsometry determines the change in polarization of reflected (or transmitted) light from a sample by measuring two parameters Ψ and Δ that characterize the relative change in the amplitudes of the p- and

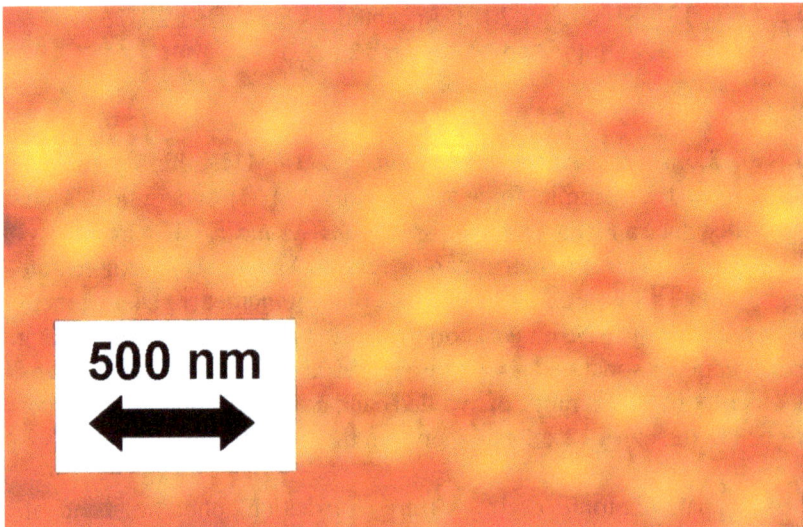

Figure 1   AFM image of the surface of self-organized opal sample

s-polarized waves and the phase shift between them:

$$\exp(i\Delta) \tan \Psi = \frac{R_p}{R_s}, \tag{1}$$

where R_p and R_s are the reflection coefficients [6]. To our knowledge, only in few cases [7–9] this method was used for characterization of opal structures. In this paper, we compare the spectra of ellipsometric parameters with conventional reflectance spectra for bulk and thin film opal samples.

2 Experimental Procedures

Thin film opals with typical sample thickness of about 10 microns were prepared on glass substrates by noise-assisted crystallization from a diluted suspension of PMMA spheres in a vertically moving meniscus [10]. Bulk opal matrices were synthesized by sedimentation of silica beads in Central Research Technology Institute "Technomash" (Moscow) [11]. AFM images were obtained by scanning probe microscope «NanoEducator» (NT-MDT, Moscow).

Angular resolved reflectance and transmittance spectra ($R(\lambda)$ and $T(\lambda)$, respectively) were measured under illumination by unpolarized white light from a tungsten lamp using a collimated beam. Reflected or transmitted light was recorded by USB650 Red Tide spectrometer (Ocean Optics, Inc.). To extract the refractive index and extinction spectra we used the spectroscopic ellipsometer "Ellips-1891" (Novosibirsk), working in the static photometric mode without any rotating elements or modulators [6].

3 Results & Discussion

The shift of the Bragg resonance wavelength λ as a function of the angle of light incidence θ obeys the combined Bragg ($2a \cos \beta = m\lambda/n$, where m is an integer) and Snell's ($n \sin \beta = \sin \theta$) laws:

$$\lambda^2 = 4a^2n^2 - 4a^2 \sin^2 \theta, \tag{2}$$

where $a = 0.816D$ is the interplane distance for (111) planes of the FCC lattice and $n = n_{eff}$ represents the effective refractive index of the opal.

One can see the standard "blue" shift of the Bragg resonance band in reflectance spectra $R(\lambda)$ of the bulk opal along the increase of the angle θ. According to our experimental results the similar shift can be observed in

$\Psi(\lambda)$ spectra (Fig. 2). In the case of better ordered PMMA thin opal films [12, 13], the dispersion of the (111) diffraction resonance deviates from the Bragg law (equation (2)) at incidence angles around $\theta \approx 51^o$ due to multiple wave diffraction at (111) and ($\bar{1}11$) or (200) planes. One can clearly observe the avoided crossing of these Bragg resonances in the interval $45° \leq \theta \leq 55°$ in reflectance and transmittance spectra (Fig. 3), where it appears in the form of double peaks.

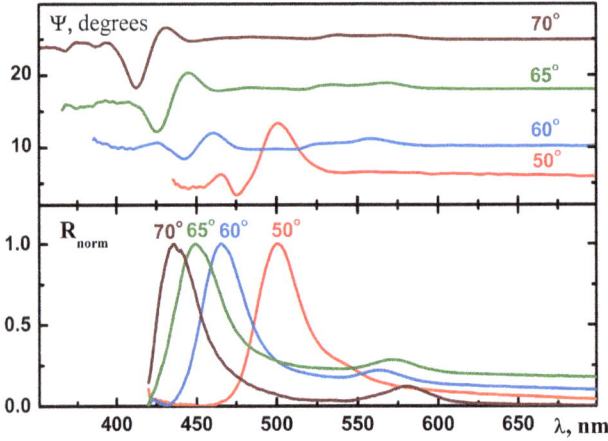

Figure 2   Ellipsometric parameter $\Psi(\lambda)$ and normalized reflectance spectra $R(\lambda)$ of bulk opal at different angles of light incidence

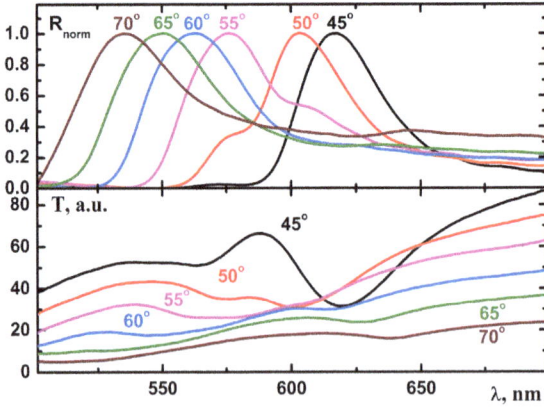

Figure 3   Normalized reflectance and transmittance spectra of the PMMA thin opal film at different angles of light incidence.

The angle dependence of the (111) Bragg resonance over the broad range of angles is linear in squared coordinates $\lambda^2 = f(\sin^2 \theta)$ with the Pearson's correlation coefficient $r = 0.999$ (Fig. 4). In this case one can use equation (2) to calculate the diameter of opal spheres $D = 315$ nm. Our previous experiments [14] have shown that the opal sphere diameters obtained from SPM-images and those calculated from Bragg reflectance spectra for opal samples agree with each other within the experimental error. The effective refractive index of $n_{eff} = 1.396$ was extracted from the Bragg fit (2) in the case of the thin opal film.

The value of refractive index n_{eff} can be also estimated using the effective medium approximation:

$$n_{eff}^2 = fn_1^2 + (1 - f)fn_2^2, \tag{3}$$

where $f = 0.26$ is the volume fraction of air voids in the FCC crystal of touching spheres, $n_1 = n_{air}$, $n_2 = 1.489$ is the refractive index of PMMA beads, so we have $n_{eff} = 1.379$ in the good agreement with the index extracted from the Bragg fit.

In the case of the bulk silica opal, the effective index of refraction, as calculated from the dispersion of the (111) Bragg resonance (eq. (2)) (Fig. 4), agrees with the ellipsometric data as well. In our bulk silica opals the porosity

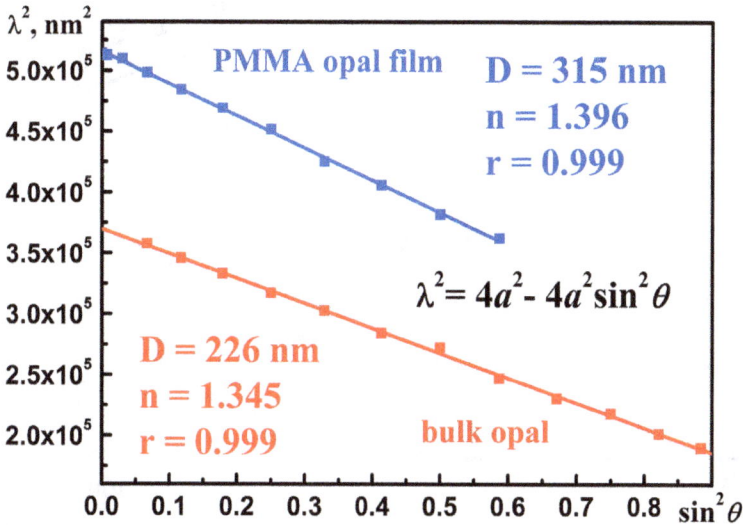

Figure 4 Angular dispersion of the (111) Bragg resonance in the PMMA thin opal film and in the bulk silica opal.

214   *Natalia Alekseeva et.al*

is $f = 0.33$, because the silica beads contain pores in their internal structure. Then one can estimate the same value of the effective refractive index from equation (3): $n_{eff} = 1.356$.

The spectra of the refractive index $n(\lambda)$ and the extinction coefficient $k(\lambda)$ (Fig. 5) of the thin opal film were calculated from the ellipsometric data. To the best of our knowledge, no interpretation of such data can be found in the current literature. We suggest the qualitative model, which takes into account the fact that light in PhC is transported by the Bloch modes. As per se, the state of linear polarization of the incident light is not preserved in the opal PhC. Calculations show that the E-vector in the Bloch mode can point in different directions for different points of the unit cell. Moreover, the field orientation rapidly changes along changes of the frequency and the quasi-wavevector of the mode [15]. This consideration makes void the standard interpretation of ellipsometric data, which is designed to homogeneous media.

Tentative explanation of $n(\lambda)$, $k(\lambda)$ spectra can be obtained with the help of the energy band structure of the opal PhC calculated using plane wave expansion method [12]. Fig. 6 shows the fragment of this structure in the range of the avoided band crossing that is relevant to experimental data. In contrast to homogeneous media, in which index of refraction is defined by

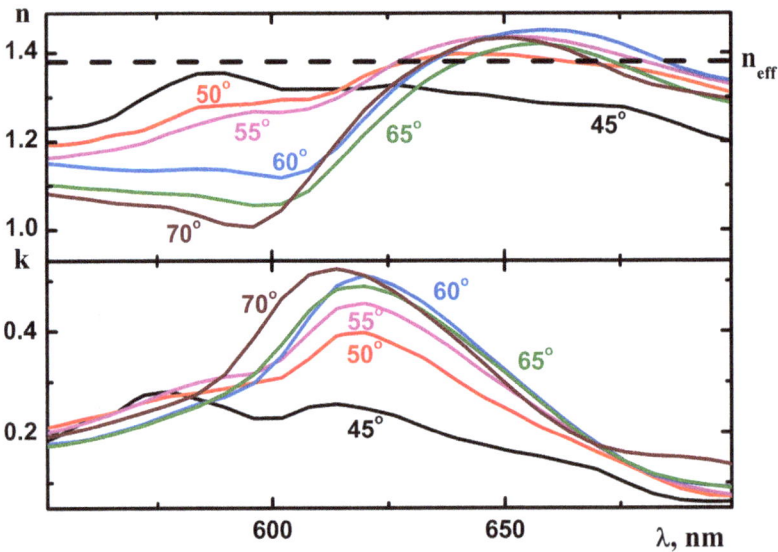

Figure 5   Refractive index $n(\lambda)$ and extinction coefficient $k(\lambda)$ of PMMA thin opal film calculated from ellipsometric data at different angles of light incidence.

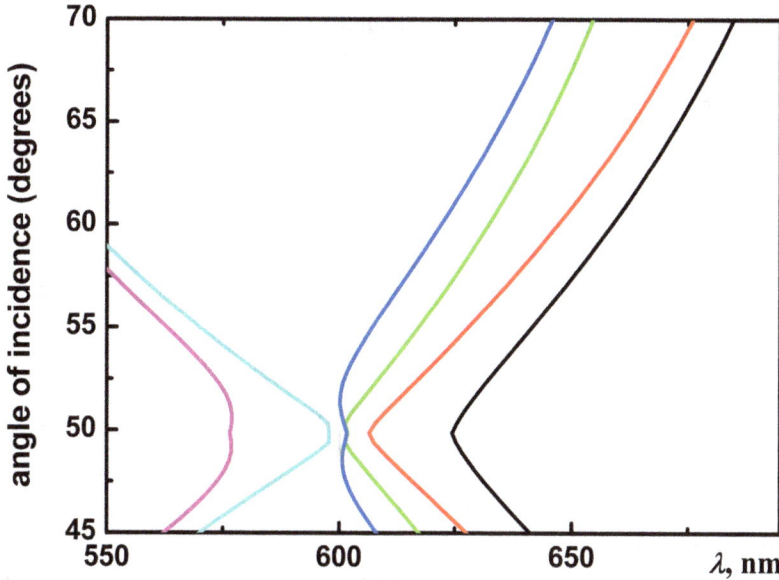

Figure 6   Fragment of the band structure of the opal film in the ΓLKL cross-section of the FCC Brillouin zone.

the polarizability of the material, in PhCs this index depends also on the light interaction with the crystal structure, so that both factors affect the mode dispersion. As the result, the group velocity and the mode index are determined by the mode dispersion. Another specific feature of 3-dimensional PhCs is that the incident light excites eigenmodes through their projections on the surface of the Brillouin zone and different eigenmodes can be coupled simultaneously. In such circumstances, the index obtained from ellipsometry data can approach the value obtained from effective medium approximation, if and only if all excited eigenmodes possess linear dispersion with similar mode index. These conditions are nearly fulfilled in the long wavelength part of the studied spectral interval, where n_{eff} closely approximates $n(\lambda)$. Accordingly, at the opposite end of the spectral range the steeper dispersion of eigenmodes leads to lower value of $n(\lambda)$.

The interpretation of the imaginary part is more straightforward. The peak of light losses centered at ~620 nm corresponds to the range of avoided band crossing, where different eigenmodes are strongly interacted each other. This interaction facilitates the efficient exchange of energy between different modes. As the result of such redistribution the light energy in the zero diffraction order is reduced. In turn, such reduction is counted by ellipsometric parameters as the increased magnitude of losses and produces a band in $k(\lambda)$ spectra. Owing to the multidirectional E-field distribution in Bloch modes such energy transfer becomes the angle-independent property.

4 Conclusion

We have shown that the ellipsometry provides complementary information to the data obtained by conventional optical characterization of opal crystals. This high-sensitive experimental technique makes it possible to discover some additional features in the wavelength- and angle-dependent optical response of these materials, especially at large angles of light incidence, near the Brewster angle θ_B ($\tan\theta_B = n_{eff}$) where the difference between s- and p- polarized electromagnetic waves becomes pronounced. The added value of the ellipsometry is that it can determine the effective index of refraction n_{eff} without assuming any material parameters [7]. As it was shown in Section 3, the obtained index value corroborates the estimates based on the angular dispersion of Bragg resonance or on the effective medium approximation in the spectral range of the linear dispersion of opal eigenmodes.

The unique observation provided by ellipsometry is the detection of the energy exchange between PhC eigenmodes in the range of avoided band crossing, which appears in linear transmission and reflectance spectra as the doubling of the diffraction resonance.

5 Acknowledgments

The authors are grateful to Prof. M. I. Samoilovich for providing high-quality bulk opal matrices. This work was supported by the Ministry of Education and Science of Russian Federation according to the program "Development of Scientific Potential of Higher Educational Institutions", by German Academic Exchange Service (DAAD) and DFG funded Cluster of Excellence "Engineering of Advanced Materials" (Germany).

References

[1] V. N. Astratov, V. N. Bogomolov, A. A. Kaplyanskii, A. V. Prokofiev, L. A. Samoilovich, S. M. Samoilovich, Yu. A. Vlasov, *Il Nuovo Cimento*, **17D**, 1349–1354 (1995).

[2] S. G. Romanov, N. Gaponik, A. Eychmüller, A. L. Rogach, V. G. Solovyev, D. N. Chigrin, C. M. Sotomayor Torres, In *Photonic crystals: Advances in design, fabrication, and characterization*; Busch K.; Lölkes S.; Wehrspohn R. B.; Föll H.; Eds.; Weinheim, DE (2004).

[3] V. Solovyev, Y. Kumzerov, S. Khanin, *Physics of regular matrix composites (Electrical and optical phenomena in nanocomposite materials based on porous dielectric matrices)*; Saarbrücken, DE, 2011 (in Russian).

[4] V. G. Balakirev, V. N. Bogomolov, V. V. Zhuravlev, Y. A. Kumzerov, V. P. Petranovskii, S. G. Romanov, L. A. Samoilovich, *Crystallography Reports*, **38**, 348–353 (1993).

[5] V. A. Tkal, N. A. Voronin, V. G. Solov'ev, N. O. Alekseeva, S. V. Pan'kova, and M. V. Yanikov, *Inorganic Materials*, **46**, 119–121 (2010).

[6] V. A. Shvets, E. V. Spesivtsev, S. V. Rykhlitskii, and N. N. Mikhailov, *Nanotechnologies in Russia*, **4**, 201–214 (2009).

[7] M. Ahles, T. Ruhl, G. P. Hellmann, H. Winkler, R. Schmechel, H. von Seggern, *Optics Communications*, **246**, 1–7 (2005).

[8] A. I. Plekhanov, V. P. Chubakov, and P. A. Chubakov, *Physics of the Solid State*, **53**, 1145–1151 (2011).

[9] A. Reza, Z. Balevicius, R. Vaisnoras, G. J. Babonas, A. Ramanavicius, *Thin Solid Films*, **519**, 2641–2644 (2011).

[10] W. Khunsin, A. Amann, G. Kocher, S. G. Romanov, S. Pullteap, H. C. Seat, E. P. O'Reilly, R. Zentel, C. M. Sotomayor Torres, *Adv. Func. Mater.*, **22**, 1812–1821 (2012).

[11] A. F. Belyanin, M. I. Samoilovich, *Nanostructures and Photon Crystals: Collective Monograph after the Materials of Plenary Reports of the 10th International Conference "High Technology in Russian Industry"*; Moscow: CRTI "Technomash" (2004).

[12] S. G. Romanov, T. Maka, C. M. Sotomayor Torres, M. Müller, R. Zentel, D. Cassagne, J. Manzanares-Martinez, and C. Jouanin, *Phys. Rev. E*, **63**, 056603 (2001).

[13] V. G. Solovyev, S. G. Romanov, D. N. Chigrin, C. M. Sotomayor Torres, *Synthetic Metals*, **139**, 601 (2003).

[14] N. Alekseeva, V. Veisman, A. Lukin, S. Pan'kova, V. Solovyev, M. Yanikov, *Nanotechnics*, **31**, 23–26 (2012) (in Russian).

[15] C. Wolff, S. G. Romanov, J. Küchenmeister, U. Peschel and K. Busch, *submitted*.

Biographies

Natalia Alekseeva finished her post-graduate course in Solid State Physics at Pskov State Pedagogical University (Russia) in 2010. Her research activity is mostly in the fields of the scanning probe microscopy (SPM) and physics of nanostructures based on porous alumina.

Grigory Cema is a third-year post-graduate student in Physics of Condensed Matter at Pskov State University (Russia). His research is focused on the study of optical characteristics of nanocomposite materials by photoluminescence and ellipsometry as well as on X-ray physics.

Aleksey Lukin is a third-year post-graduate student in Solid State Physics at Pskov State University (Russia). His research is focused on the electrical characterisation of zeolite-based nanocomposites as well as on the preparation of opal-based metal-dielectric systems.

Svetlana Pan'kova graduated in Physics and Mathematics in 1987 and got her Ph.D. in Solid State Physics at Herzen State Pedagogical University of Russia (St. Petersburg) in 1998. Today Svetlana Pan'kova is Associate Professor at Pskov State University. Her research activity is mostly in the fields of the scanning probe microscopy and electrical characterisation of opal-based nanostructures.

Sergei Romanov received his Diploma from the Polytechnical Institute of Leningrad, USSR in 1978. He obtained PhD and DSc degrees from the Ioffe Institute in 1986 and 2013. Since 1990 he is a research professor at Ioffe Institute. He authored a number of pioneering results in the physics of regular ensembles of templated nanostructures. He also promoted this approach to realization of low-dimensional materials working at universities in Glasgow, Wuppertal, Cork and Erlangen. He currently focused on designing complex photonic and plasmonic architectures based on colloidal platforms.

Vladimir Solovyev got his diploma from Pskov State Pedagogical Institute (USSR) in 1976, his PhD and DSc degrees from Herzen State Pedagogical University of Russia (St. Petersburg) in 1991 and 2005. Today he is a Professor, Head of Physics Department at Pskov State University. He is the author of about 50 regular papers and 3 monographs. Major research interests focus around electrical and optical phenomena in nanocomposite materials based on regular porous dielectric matrices.

Victor Veisman got his diploma from Pskov State Pedagogical Institute (USSR). Today he is Associate Professor at Pskov State University. He is the author of more than 100 regular papers, supervisor of dozens Diploma students, experienced lecturer on optics, atomic and solid state physics, quantum mechanics and astronomy. His research interests are in electrical and optical properties of point defects in alkali halide crystals, as well as in methods of preparation of nanocomposites based on regular porous dielectric matrices.

Mikhail Yanikov finished his post-graduate course in Solid State Physics at Pskov State Pedagogical University (Russia) in 2008. His research activity is mostly in the field of reflectance and transmission angle-resolved optical spectroscopy of opal-based 3-dimensional photonic crystals.

Online Manuscript Submission

The link for submission is: www.riverpublishers.com/journal

Authors and reviewers can easily set up an account and log in to submit or review papers.

Submission formats for manuscripts: LaTeX, Word, WordPerfect, RTF, TXT.
Submission formats for figures: EPS, TIFF, GIF, JPEG, PPT and Postscript.

LaTeX

For submission in LaTeX, River Publishers has developed a River stylefile, which can be downloaded from http://riverpublishers.com/river publishers/authors.php

Guidelines for Manuscripts

Please use the Authors' Guidelines for the preparation of manuscripts, which can be downloaded from http://riverpublishers.com/river publishers/authors.php

In case of difficulties while submitting or other inquiries, please get in touch with us by clicking CONTACT on the journal's site or sending an e-mail to: info@riverpublishers.com

www.ingramcontent.com/pod-product-compliance
Lightning Source LLC
Chambersburg PA
CBHW061839220326
41599CB00027B/5342